Introductory Chemistry Modules: A Guided-Inquiry Approach

Joe L. March, *University of Alabama – Birmingham*

K. Kenneth Caswell, *University of South Florida*

Jennifer Lewis, *University of South Florida*

Contributions by Craig McClure, *University of Alabama – Birmingham*

HOUGHTON MIFFLIN COMPANY   BOSTON   NEW YORK

CHAPTER 1

Jumping Through Science with a P.O.G.I.L. Stick

ACTIVITY 1.1

Objectives

- Understand the approach taken by the course.

- Recognize how your course grade will be determined.

The Model

Refer to your syllabus and this book.

Reviewing the Model

1. Will exams be given in this course? If so, how many? When is the first exam?

2. Use the Table of Contents of this book and the syllabus to create a list of topics that are covered in the first five weeks this course.

3. What activities are graded in this course?

4. What are the required materials (textbooks, calculators, study guides, notebooks, other) for this course?

Exploring the Model

5. What activity in this course has the greatest effect on your final letter grade?

Exercising your Knowledge

6. Why is each member of your group taking this chemistry course?

7. Review the syllabus and this book and explain how the course is structured for you to be successful in this course and future chemistry classes.

8. An important aspect of this book involves asking you to talk with your classmates. Thus it is important to know your classmates. List each group member's name and one fact about them to help you get to know them better.

Summarizing Your Thoughts

9. As a group, discuss which one of the activities associated with this course is likely to prove most essential to understanding and achieving your goals in chemistry?

10. Focusing on a grade in a course is a short-term goal. As a group, identify at least three long-term goals for your introduction to chemistry.

ACTIVITY 1.2

Getting Started

The scientific method is a process for forming and testing solutions to problems or theorizing about how or why things work. It reduces the bias of the researcher and involves experimentation. Once started, the scientific method cycles over and over again to further refine an initial hypothesis.

Objectives

- Identify the steps of the scientific method

- Be able to formulate a hypothesis to test experiments in different situations

- Understand how the scientific method is applicable to everyday-life situations

- Be able to identify different approaches to the scientific method

Model

Figure 1.1

Allie's car didn't start this morning. She thought her battery was dead, so she charged it while she was at school. When she came home the battery was charged, so she tried to start the car. It still didn't work. She called her dad to tell him what was going on. He suggested that something else was wrong with her car.

As her initial hypothesis was incorrect, she thought it might be her starter that wasn't working—*that* was what was wrong with her car.

Reviewing the Model

1. What was Allie's problem?

2. What was Allie's initial reaction to the problem?

3. What did she do next?

4. What was her first conclusion?

5. What was the solution to Allie's problem?

Exploring the Model

6. Identify the term "hypothesis" in the Model, and work as a group to create a definition for hypothesis in your own words.

7. Identify all the steps Allie took to solve her problem.

8. Fill in the table using information from the Model and no. 7.

Table 1.1   **The Scientific Method**

Steps of the Scientific Method	Actions taken by Allie
	Allie needed to get to school, but her car would not start.
Hypothesis	
Experimental Test	
	A charged battery did not crank the car.
Report Results	

a.   The method is shown in order here. Could you start with someone else's report results?

b.   Does the method have to be performed in this order? Why or why not?

c.   Can an experimental test lead to more than one alternative hypothesis?

Exercising Your Knowledge

9.   Create a story about a situation that is familiar to you that requires you to collect some information and make a decision. Your story should contain all of the aspects of the scientific method you identified.

a.   Use Table 1.2 to organize your story and to clarify how you have related it to the scientific method.

Table 1.2   **Organizing Your Story**

Steps of the Scientific Method	Actions in Your Story

Summarizing Your Thoughts

10. Explain how you use the scientific method in your everyday life without actually realizing you are doing so.

11. Draw your own conclusions about why the scientific method is important to chemistry.

12. What new information did you learn today that will add to your previous knowledge of the scientific method?

13. Describe how the scientific method might be useful when approaching textbook problems in any course.

ACTIVITY 1.3

Getting Started

Success in your studies is ultimately your responsibility. Instructors have an important role, but in any classroom model the level of knowledge you acquire will depend on your ability to personally work through the key concepts of the material.

Still, you won't have to go through this course alone. We will be utilizing the talents and abilities of your classmates throughout the semester. However, for this to be successful everyone has to take some responsibility. There are many ways to help people work together, and your instructor may require you to establish formal roles (leader, recorder, reflector, or similar roles). However, this activity is designed to help you think about some very basic responsibilities related to working with your classmates.

The Model

Contract For Effective Group Work

- All members of the group will contribute to each and every activity.

- If a member of the group does not understand the material, the members of the group that do understand will work to explain the material to their classmate.

- Each group member shall be free to express their opinion without fear of ridicule from any of the other group members.

- Each group member will do everything in their power to attend every class so that they may help their classmates and themselves learn the material.

- Each member of the group is responsible for making sure that the other members of the group are working and staying on task. They are also responsible for making sure that no one person dominates the discussion.

Signed

The Members of the Group

Reviewing the Model

1. How many members of the group are responsible for contributing to all of the activities?

2. Who is responsible for signing the contract for Effective Group Work?

Exploring the Model

3. Think of a time when working in a group is effective (sports, emergency responders, military forces). Do these groups have an implied group agreement?

4.   Select an example of an effective group, and write a "contract for effective group work" for that group.

5.   How do groups hold members responsible for behaviors identified in a contract?

6.   Prepare a list of behaviors that ideal students exhibit when working in a group.

7.   The contract states that all members of the group are *responsible for making sure the workload is distributed evenly*. How can you make sure this is happening?

8.   What actions should be taken if a group member decides to miss a class?

9.   Many times a member of a group will not ask a question about something they do not understand to avoid being made fun of or feeling less intelligent than the other members of the group. What steps can you take as a group to make everyone feel comfortable enough to ask a question? Are there other ways to assure that everyone understands the topics that are being discussed?

Summarizing your Thoughts

10.   As a group, write your own contract. Be sure to list in your contract what actions should be taken if members of your group fail to live up to the contract.

CHAPTER 2

Representing the Little Things in Life

Getting Started:

One of the challenges in chemistry is that we must possess the ability to think about chemistry from four different perspectives: a macroscopic view, a descriptive vocabulary, a symbolic representation, and a nanoscale view. This ability is similar to learning four different languages. You have to be fluent in all four and able to think about how the other three perspectives might be useful for every chemical process. This chapter will introduce the four perspectives and explore how chemists use these different ways of thinking.

Chapter Goals:

- Become familiar with common symbols and their use

- Use symbols to represent descriptive vocabulary words or pictures

- Recognize a nanoscale picture and be able to represent it with a symbol or descriptive vocabulary

ACTIVITY 2.1

Objective

- Recognize the importance of symbols or abbreviations
- Identify an approach to understanding unfamiliar abbreviations

Getting Started

Every day we use symbols, words, and drawings in various ways to show the same thing. Our initial model begins with common abbreviations and symbols so that we can begin to think about the importance of these symbols or abbreviations.

The Model:

Table 2.1

Macroscopic View	Descriptive Vocabulary	Symbolic Representation
	Dollar	$
	Penny	¢

Reviewing the Model

1. What is the symbolic representation for Dollar? _____

2. What is the descriptive vocabulary word that is represented by the symbol ¢? _____

Exploring the Model

3. Both the dollar and the penny represent money. How do you decide which symbol to use when writing an amount?

4. Do you visualize the macroscopic dollar when you see the $ symbol? _____

5. Are all symbols simply an abbreviation of the vocabulary word or phrase? _____

6. Prepare a list of at least three other everyday quantities that use symbols to represent the unit of measure.

7.   Provide an example of a symbol that is not simply an abbreviation of the vocabulary word.

8.   Why would you choose to use symbols instead of the entire word?

9.   Where or when do symbols cause confusion?

Exercising Your Understanding

10.   Fill in the missing parts of the table.

Table 2.2

Macroscopic View	Descriptive Vocabulary	Symbolic Representation
	Liter	
		Tsp.

11.   Classify each symbol in relation to its relative size:

   a.   in. (small)   mi. (large)

   b.   lb. _____   ton _____

   c.   yr. _____   s. _____

Summarizing Your Thoughts

12.   Write a short paragraph that explains the relationships among the macroscopic view, the descriptive vocabulary, and the symbolic representation.

13.   Describe how visualizing the macroscopic view could be helpful when calculating any new quantity.

14.   What do you do when you see a new symbol that you do not understand?

15.   Was your group useful in helping everyone recognize the symbols presented? What could be done to improve the quality of the group work as you move into the next section?

ACTIVITY 2.2

Objective

- Recognize the symbols of the periodic table

Getting Started

The periodic table is an extremely valuable tool for the chemist. We will explore the organization of the table and the properties of the symbolically represented elements in more detail later. However, for now we want to recognize that the periodic table is an arrangement of the symbols that we use in chemistry to represent the elements. We use the names of the elements many times in our daily lives. We have heard of oxygen for the air we breathe and potassium as an important part of our diet. You will learn what it means to be an element and the properties of elements in detail later, but initially we will need to learn how to quickly interchange an element's name and symbol so that you can have a reasonable understanding of more complex ideas. Explore the Model to see the common traits among the symbols in this table.

The Model

Figure 2.1   The Periodic Table, Main Group and Transition Elements

Reviewing the Model

1. How many elements' symbols contain just one letter? _____

2. If there are 109 elements, how many elements' symbols contain two letters? _____

Exploring the Model

3. What case (upper or lower) is the first letter of every element? _____

4. For the elements with more than one letter, what case (upper or lower) is the second letter of every symbol? _____

5. What is the maximum number of letters that we can use to represent an element? _____

6.   From the Table, what is the most common number of letters used to represent an element? _____

Exercising Your Understanding

7.   Table 2.3 contains the 20 most common elements found on earth. You will encounter these elements throughout your studies and in your daily life; thus it is useful to learn to relate the names and symbols for these elements. Use a periodic table to complete the table:

Table 2.3 The Twenty Most Common Elements on Earth

Element	Symbol
Hydrogen	
	He
Neon	
Fluorine	
Oxygen	
	S
Carbon	
	N
Aluminum	
Sodium	
Potassium	
Iron	
	Ni
Copper	
Iodine	
	Cl
	Pb
Silicon	
Bromine	
Magnesium	

8.   Some elements' symbols are derived from Latin or from the discoverer's native language. Identify the elements in Table 2.3 whose symbol does not appear to be related to the name of the element.

9.   Write the symbol for all the elements that have a symbol that begins with the letter A. Is there a simple way to associate the name with the second letter in each of these symbols?

10.   Explain why writing the symbol SN for tin (Sn) would cause confusion.

Summarizing Your Thoughts

11. After your facilitator reviews your completed Table 2.3, prepare a statement that helps you remember the rules for capitalization of the letters in an element's symbol.

12. Explain why writing a chemical symbol correctly is important when communicating chemistry to another person.

13. What would be some likely places to find a periodic table if you needed one to look up symbols of the elements?

14. What insight have you gained as a result of your team's performance today?

ACTIVITY 2.3

Objectives

- Use chemical symbols from the periodic table to represent molecular formulas

- Recognize a nanoscale picture and be able to represent it with a symbol or descriptive vocabulary

Getting Started

The next level representation used in chemistry is a chemical formula. A chemical formula represents an individual molecule or compound. A compound is any group of atoms that are bound together in a regular pattern. A molecule is the smallest particle that can be obtained with the same characteristics as the bulk compound. If we break up a molecule, we can end up with atoms, but atoms have different characteristics than the molecule. There are an infinite number of chemical formulas made up from our 109 elements. There are different types of chemical formulas that we will introduce later. First we will explore molecular formulas.

Molecules and atoms are much too small for us to be able to see, and if we were able to see them, they would most certainly not look like any of these figures. The different representations of a molecule help us understand how molecules behave and provide us with a way to understand and visualize their behaviors. Through the years, our ideas of what molecules may look like have advanced, but it is still impossible to accurately represent what molecules look like. So we rely on chemical symbols or models that we *can* examine to try to understand why molecules behave as they do.

We start with the water molecule. There are many different representations of a molecule of water, and each shows a different characteristic of water. We will probe these representations to see the value of the different representations and then practice interchanging the nanoscale representation with the symbolic.

The Model

Table 2.4 Relating The Nanoscale to Molecular Formulas

Ball-and-Stick Model	Molecular Formula	Atoms of First Element	Atoms of Second Element
	CH_4	1	4
	H_2O	2	1
	NH_3	1	3
	HCl	1	1

Reviewing the Model

1.   What is the shape used most often to represent the elements? _____

2.   How do we show that elements are connected in the Model?

3.   Do all of the elements have the same size? _____

4.   Sketch one of the representations of water, and label the oxygen and hydrogen.

5.   Is the hydrogen atom in this table being represented by a white or a grey sphere?

6.   How many hydrogen atoms in

     a.   HCl _____

     b. NH_3 _____

     c. CH_4 _____

     d. H_2O _____

Exploring the Model

7. Look at the three representations of water. Determine which drawing is most useful for describing:

    a.   the volume of the molecule.

    b.   the H-O-H angle.

    c.   the connections (bonds) between each atom.

8. How does the subscript indicate the number of each element in each compound?

9. Is a subscript used when there is only one atom of an element in the compound?

10. Does the hydrogen atom appear in the same place in all of the chemical formulas from the table above?

11. All of the compounds in the Model are colorless gases. Would it be possible to tell them apart on the basis of color?

Exercising Your Understanding

12. Give the number of atoms of each element present in one molecule of each of the following compounds.

Table 2.5 Interpreting the Subscript

Compound Name	Molecular Formula	Atoms of First Element	Atoms of Second Element
Dihydrogen sulfide	H_2S		
Dinitrogen tetroxide	N_2O_4		
Carbon dioxide	CO_2		
Carbon monoxide	CO		
Carbon tetrafluoride	CF_4		

13. Complete Table 2.6 by writing the molecular formula for each compound.

Table 2.6   Writing Molecular Formulas

Compound Name	Molecular Formula	Atoms of First Element	Atoms of Second Element
Nitrogen monoxide		1	1
Sulfur dichloride		1	2
Sulfur trioxide		1	3
Boron tribromide		1	3
Dicarbon hexahydride		2	6
Diphosphorus Pentoxide		2	5

14. Draw a nanoscale representation of sulfur dioxide and sulfur trioxide. Make sure your drawings are clearly labeled.

15. In your drawings, how did you decide what connections to make?

16. What other information would have been useful to know when you started to draw SO_2 and SO_3?

17. Would you expect sulfur dioxide and sulfur trioxide to have the same properties? Explain your answer.

Summarizing Your Thoughts

18. What can we say about what a chemical formula tells us about the ratio of one element to another? Use an example from Table 2.6 to justify your answer.

19. We saw examples in which we used prefixes like "mono" or "di" to represent one or two elements, respectively, in a chemical formula. Complete the following Table for use later when we explore nomenclature further.

Table 2.7 Prefixes That Describe the Number of Atoms in a Molecular Formula

Number of Atoms of Element	Prefix Used
One	
Two	
Three	
Four	
Five	

20. In everyday use, we encounter both water and alcohol. We know that both liquids are clear, colorless liquids. However, when consumed we know that they have very different effects. Compare the ability of the three representations (nanoscale, symbolic, and macroscale) to predict physiological effects. Which representation is most useful? Which representation is least useful? Which representation is easiest to include in a written communication?

CHAPTER 3

So You Want to Mix Things Up? (States of Matter and Mixtures)

Chapter Goals:

- Describe the three states of matter (solid, liquid, and gas) from a nanoscale perspective

- Understand how the placement on the periodic table can help identify the state of matter

- Show a symbolic representation of a mixture on the nanoscale

- Make a determination between two types of mixtures

ACTIVITY 3.1

Objective

- Identify the three states of matter (solid, liquid, gas) from a nanoscale perspective

Getting Started

We are very familiar with the three states of matter. In our everyday lives we come into contact with all three forms: solid, liquid, and gas. Water is the most familiar example. We have seen water as steam from a teapot, as a liquid in a swimming pool, and as a solid on a hockey rink. Water's amazing ability to exist in all of these forms within the small temperature range of our everyday life makes it very special in the chemical world.

Because of its familiarity to you on the macroscopic scale, we will examine the nanoscale view of the three states using water as our model. Recall that water is written as H_2O for the chemist. We will represent the atoms as spheres as we work through this model to gain some understanding of the nanoscale view of solids, liquids, and gases.

The Model

Table 3.1 Three Views of Water

	Gas	Liquid	Solid
Looking at Water	Molecules are very far apart	Molecules are closer together but still very separated	Molecules are very close together and densely packed
Shape	Will take the shape of the container	Will take the shape of the container	Will form its own shape and be rigid or fixed
Volume	Takes the volume of the container	Has a constant volume independent of the size of the container	Has a constant volume independent of the size of the container
Particle Motion	Random motion throughout the container	Random motion throughout the container	Local vibration in a fixed position, no long-range motion

Reviewing the Model

1. In which state of matter are the water molecules farthest apart?

2.   What is meant by *local vibration* in the description of the particle motion of a solid?

3.   What state would you expect to find water in at room temperature?

Exploring the Model

4.   Consider how the three states of matter relate to their container. Describe what you would expect to see in the macroscopic view if you were to try to fill a balloon with each state of matter.

5.   Do your macroscopic expectations of filling a balloon agree with the Model's description of shape?

6.   Imagine you fill a piston (or a syringe) with a gas. Would you expect to be able to move the piston?

7.   Would you be able to move the piston if it were filled with rocks?

Summarizing Your Thoughts

8.   Explain how the relative positions (how close they are to each other) of the molecules or atoms at the nanoscale affect your ability to move the piston.

9.   Players on a soccer team could be found in three places: 1) on the field playing, 2) on the bench sitting with their team, or 3) back at their respective homes. Explain which state of matter could be described by each situation.

10.   Are there problems with the analogy of the soccer team to the states of matter? Explain where this analogy might not be appropriate.

ACTIVITY 3.2

Objective

- Understand how placement on the periodic table can help identify the state of matter of an element under normal conditions

Getting Started

The periodic table can be a very valuable tool for the chemist. In chapter two we saw how each block represents a different element. We will continue to explore how we can continue to gain more insight from the periodic table as we proceed, but for now let us take a look at how those elemental symbols are arranged on the periodic table to help identify different characteristics of the element.

The Model

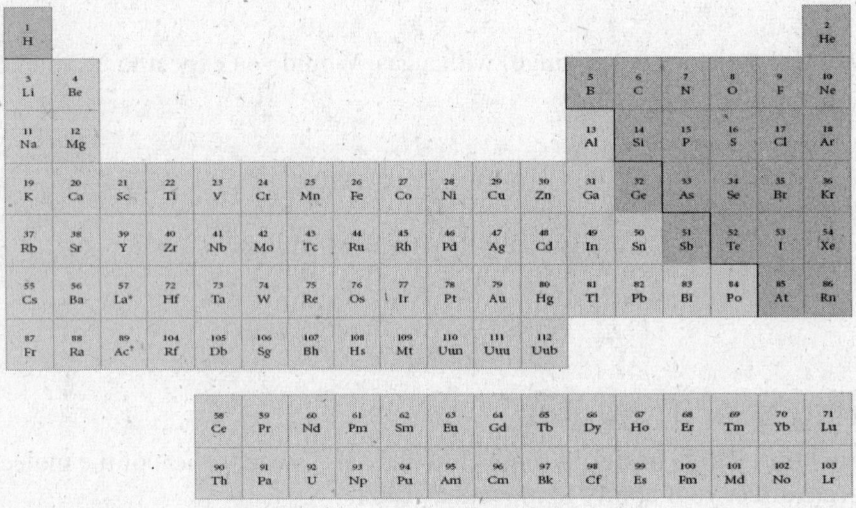

Figure 3.1   The Periodic Table

Reviewing the Model

1.   Where is the element iron (Fe) located on the periodic table?

2.   Where is the element oxygen (O) located on the periodic table?

3.   Where is the element aluminum (Al) located on the periodic table?

Exploring the Model

4.   Describe the macroscopic properties of iron (Fe).

5.   Describe the macroscopic properties of elemental oxygen (O_2) as found in the air.

6.   Review the elements around iron on the periodic table. Are their macroscopic properties similar to iron?

7.   Review the elements around oxygen on the table. What do you know about their macroscopic properties?

8.   What shading is being used to describe nonmetals?

9.   What shading is being used to describe metals?

Exercising Your Knowledge

10.   How close (or how far away) do you expect to find molecules of chlorine (Cl_2)? Why?

11.   You have not likely ever seen elemental sodium, but based on the position of sodium on the periodic table would you expect it to appear more like iron or like oxygen? Explain your choice.

Summarizing Your Thoughts

12.   Write yourself a note that describes the side of the periodic table where metals are likely to be found.

13.   Where would you expect to find elements that don't appear as metals?

ACTIVITY 3.3

Objectives

- Show a symbolic representation of a mixture on the nanoscale
- Make a determination between two types of mixtures

Getting Started

Now that we understand the different types of matter, and we are familiar with how the periodic table can help us to understand what type of material we have, we can begin to look at what will occur when we combine compounds that are in different states.

The Model

A student is given two beakers of water. In the first beaker he adds a large scoop of sugar, and in the second he adds a large scoop of sand. He stirs them up for a minute and then reports what he sees.

Sugar beaker: He could no longer see the sugar, and the beaker of water looked clear.

Sand beaker: The sand settled in the bottom of the beaker.

Bowl of sugar   Bucket of Sand

Figure 3.2   Sugar and Sand

Table 3.1

Macroscale	Descriptive Vocabulary	Formula of the Compounds	Nanoscale	Key
	Sugar and Water	$C_6H_{12}O_6$ & H_2O		water / sugar
	Sand and Water	SiO_2 & H_2O		SiO_2 (simplified representation)

Reviewing the Model

1. Are the sugar molecules distributed within the water?

2. Is the SiO_2 distributed in the same manner?

Exploring the Model

3. Describe the macroscopic differences between the two mixtures.

4. Describe how the nanoscale pictures differ.

Exercising Your Knowledge

5. After observing the beakers the student was instructed to take a cup and remove some water from the top portion of the sugar beaker. He then took a cup and removed some water from the top portion of the sand beaker. Sketch a nanoscale view of the liquid that was removed from each beaker.

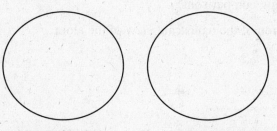

     Sugar Beaker   Sand Beaker

6. Based on what you have learned, predict whether the following mixtures would be <u>heterogeneous</u> or <u>homogeneous</u>:

    Orange juice (with no pulp)   _____

    Clear ocean water   _____

    A piece of raisin bread   _____

    The air inside a helium balloon   _____

    A pond with some algae   _____

7. Explain how thinking about the nanoscale view of a mixture could be helpful in describing a situation where a solid material appears in your glass of water.

8. In no more than three sentences summarize what was learned about mixtures today.

CHAPTER 4

How Many Gills in a Kilomutchkin? (Measurements and Units)

Chapter Goals:

- Recognize the importance of units and be able to identify various common units

- Make rational choices about which units to use

- Identify equalities, and use them in conversion factors

- Understand standard SI units, and memorize the meanings of common prefixes

- Use prefixes to aid conversions within the standard SI unit system

- Understand the value of unit conversions in allowing comparisons

- Through the use of SI units, gain greater appreciation of the nanoscale view of the atom

ACTIVITY 4.1

Objective

• Recognize the importance of units, and identify common units

Getting Started

As chemists we deal with numbers every day. For example, numbers are a useful way for a forensic chemist to describe a toxic dose of arsenic or for an industrial chemist to describe how much citrus peel will be needed to meet customer demand for sweet-smelling dog shampoo. Since chemists so often communicate with numbers, they have developed standards for communication that help prevent misunderstandings. Fortunately, as you'll discover in this activity, these standards are "common sense," and you probably use the same ideas in your daily life.

The Model

You are driving to find a friend's house and get lost. You stop and ask a stranger walking his dog for directions, and he seems to know where you are going. He says:

"Take a left at the next stoplight, and then go five."

Reviewing the Model

1.   What direction did the man tell you to turn? _____

2.   How far did the man tell you to drive? _____

Exploring the Model

3.   What is it about these directions that might cause confusion?

4.   What question would you ask of the man in order to clarify these directions?

Exercising Your Knowledge

5.   Complete the table using units of measurement that members of your group might use every day.

Table 4.1

Types of Measurement:	Common Units of Measure:		
Length of a pencil	Inches	or	Centimeters
Length of a soccer field		or	
Time spent studying chemistry		or	
Time spent in high school		or	

6.   Is "5 pounds" more likely to refer to the mass of a pumpkin or the length of a French fry?

7.   Is "2 teaspoons" more likely to refer to the amount of sugar in a cup of coffee or the amount of corn grown in Iowa annually?

8.   What factors should you consider when choosing which unit of measurement to use in a particular situation?

Summarizing Your Thoughts

9.   Briefly explain, using the examples you listed in the Table, why it would be necessary to include the units when referring to time.

10.   Why is including the unit of measurement important when communicating with numbers?

11.   What do chemists need to think about when choosing which unit of measurement to use?

ACTIVITY 4.2

Objective

- Identify equalities, and use them in conversion factors

Getting Started

From the previous activity we have seen that it is necessary to include the unit of measurement when trying to communicate with someone about a measurement. But what do you do if people are unfamiliar with the unit of measurement you want to use? We will now see how some units of measurement can be related to each other through the use of an *equality*, or *conversion factor*.

An equality is a mathematical statement of the relationship between two different units of measurement. A conversion factor is a rearranged form of the equality that can be used to convert one unit of measurement in the equality to the other by simple multiplication. We start with an unfamiliar equality so you can focus on how you process this information in order to create a conversion unit.

The Model

A pastry chef needs 12 *firkins* of Key lime juice to make pies for a banquet, but Key lime juice is sold only in *pins*.

The equality: 2 pins = 1 firkin

The conversion factor for converting pins to firkins: $\dfrac{1\,firkin}{2\,pins}$

The conversion factor for converting firkins to pins: $\dfrac{2\,pins}{1\,firkin}$

Reviewing the Model

1.   How many pins are in a firkin? _____

2.   How many pins are in 12 firkins? _____

3.   Write the conversion factor you used to determine how many pins are in 12 firkins.

Exploring the Model

4.   What is the relationship between the conversion factor for converting pins to firkins and the conversion factor for converting firkins to pins?

5.   How is each conversation factor related to the equality?

6.   Is $\dfrac{1\,firkin}{2\,pins}$ equal to $\dfrac{2\,pins}{1\,firkin}$? Explain your answer.

7.   Is $\dfrac{1\,firkin}{2\,pins}$ equal to one? Explain your answer.

8.   Use your answers to numbers 6 and 7 above to explain why conversion factors can be used to convert one unit of measure to another with simple multiplication.

Exercising Your Knowledge

9.   There are also 4 gills in 1 mutchkin, and 2 mutchkins in 1 chopin. Express this statement as an equality.

10.   What is the conversion factor for gills to mutchkins?

11.   What is the conversion factor for mutchkins to gills?

12.   If you have 6 mutchkins, how many gills do you have?

13.   There are 2 mutchkins in 1 chopin. If you have 12 gills, how many chopins do you have? Show your work, and label all conversion factors.

> **Some Common Equalities:**
>
> 1 inch is the same as 2.54 cm
> 1 pound is equal to 16 ounces
> 1 ton = 2,000 pounds
> 1 gallon is the same as 4 quarts
> There are 60 seconds in one minute

14.   From the above table of common equalities, list the key words or symbols that are used when identifying an equality.

15. Estimate without a calculator whether the number of seconds in 32 minutes is greater than or less than 500. What would you do if the answer shown on your calculator didn't agree with your estimate?

16. Explicitly show how conversion factors are used to calculate the number of seconds in 10 minutes.

17. If I give you 2 pounds of silver, how many ounces will you have?

Summarizing Your Thoughts

18. Describe how any equality can be turned into a conversion factor.

19. Explain how the equalities and the resulting conversion factors can help you estimate answers before you begin using a calculator.

20. Review the common units you prepared in Activity 4.1 (Table 4.1). Discuss these units with your group, and see if you can write the equalities and conversion factors that are related to the units you chose. If you are not certain of an equality, where do you look for the required information?

21. The tool you have been using to solve problems properly (in paying attention to units) is called dimensional analysis. In your own words, give a definition for dimensional analysis.

22. Briefly describe how well your group is working together. Be sure to include a specific example of a contribution made by each member of your group. (Understand that your instructor will compare the answers you give with what they observed in class.)

ACTIVITY 4.3

Objective

- Become familiar with the SI units of measurement

- Explore the use of prefixes in the SI system

- Understand the value of unit conversions in allowing comparisons

Getting Started

In scientific inquiry, units from the SI system (International System of Units) or metric system primarily are used for measurement. This standardized set of units provides everyone with a common, fundamental measurement system. In this way, we can all communicate easily with each other.

Throughout science, we work with very big and very small measured values. A value is cumbersome when we use all the decimal places and do not know how accurate it is. Therefore, we use a system of prefixes to communicate a common multiplication factor.

In this activity, we will introduce the common SI units and the multiplication factors that you will likely encounter in other science courses. In addition, we hope that through the use of these SI units you will gain a greater appreciation for the magnitude of the nanoscale.

The Model

Smoke rises from the tents of the market place of Chemlandia. Anni asks the vendor to trade her ten small gemstones for the local currency, petreekas. She knows that the current market price for her gems is 1 petreeka per gram of gemstone. The vendor tells her that he will pay her 100 millipetreekas for each gram. Anni decides this isn't enough and leaves to search for another exchange house.

Table 4.2 SI Units and Some Equalities

Measured Quantity	SI Unit (and symbol)	Equalities with Other Units
Length	meter (m)	1 meter = 3.2808 feet
Volume	liter (L)	$1 \text{ m}^3 = 264.2$ gallons 1 liter = 0.2642 gallons
Time	second (s)	
Mass	gram (g)	1 kg = 2.2046 pounds
Temperature	Kelvin (K)	$K = °C + 273.15$ $°C = (°F - 32) * (5/9)$

Table 4.3 contains the most common prefixes used in the SI system. There are others listed in your textbook; however, these should be memorized.

Table 4.3

Prefix	Units	Symbol	Verbal Description
kilo	1000	k	One thousand units
deca	10	D	Ten units
	1		One unit
deci	0.1	d	One tenth of a unit
denti	0.01	c	One hundredth of a unit
milli	0.001	m	One thousandth of a unit
nano	0.000001	n	One-millionth of a unit

Reviewing the Model

1. What is the unit of currency in Chemlandia? _____

2. What is the multiplier for milli-? _____

Exploring the Model

3. What was the vendor's offer, in units of petreekas per gram?

4. Why did Anni decide to look for another vendor?

5. Looking at Table 4.3, is centi- less than or more than one unit?

6. Would kilo- most likely be used to represent a large amount of something or a very small amount of something?

7. Explain why the name of the insect *millipede* doesn't follow SI conventions.

Exercising Your Knowledge

8.   Write down four different values for the SI unit of length by adding prefixes to the base unit.

9.   Write down four different values for the SI unit of mass by adding prefixes to the base unit.

10.   Anni determines the mass of her gemstones and finds the mass to be 7,532,000 centigrams. What would be a more appropriate unit for this measurement? Justify your answer.

11.   Using the proper equality from Table 4.2, calculate Anni's mass to kilograms.

12.   Using words only (no numbers), write the steps necessary to show how many kilograms are in a milligram.

ACTIVITY 4.4

Objectives

- Through the use of SI units, gain greater appreciation of the nanoscale view of the atom

Getting Started

In order to make direct comparisons, you must ensure that all things being compared are in the same units. Now that you are capable of converting to nanometers, you can make some comparisons.

The Model

A sulfur atom with a radius of 0.1 nm is placed in the center of a basketball that has a diameter of 35 cm.

Reviewing the Model

1. What is the radius of the basketball?

2. What difference (in factors of ten) is there between the units of the basketball and the sulfur atom?

3. What procedure would you use to determine the distance from one edge of the sulfur atom to the surface of the basketball?

Exploring the Model

4. What is the basketball's radius in nanometers?

5. In grammatically correct sentences, explain each step of the calculation required to convert the basketball's radius from cm to nm.

6. Calculate the distance between the edge of the sulfur atom and the surface of the basketball.

7.   The average distance between the earth and the moon is 384,403 kilometers. Express this distance in meters and nanometers.

8.   Compare the distance you calculated in question 6 with the distance between the earth and the moon you calculated in question 7.

9.   Anni walked 30 kilometers and did not find a favorable exchange rate with the first vendor on the edge of the market place. She walked an additional 300 meters and found a vendor willing to give her an exchange rate just over market value. In grammatically correct sentences, compare the two distances and describe why her extra travel was worthwhile.

Summarizing Your Thoughts

10.   Using Anni's experience in the market, explain how the SI prefixes make converting between two quantities easier.

11.   Wanting to know what you are learning in school, your five-year-old brother asks you, "How small is a sulfur atom?" Write a short paragraph about the size of a sulfur atom in language that your little brother would understand (Hint: use comparisons).

12.   In your own words, explain what you must know before you can convert between two measured values.

13. What information in today's activity is important to memorize?

14. In complete sentences and proper grammar, describe how equalities may be used to convert measurements among different units.

CHAPTER 5

It's a Small, Small World (Atomic Theory)

Getting Started

How do we describe something we cannot see directly with our eyes? Very often we turn to drawings or models that are familiar to us. Our use of the sphere to represent atoms allows us to think of atoms as balls. This approach works for some observations, but we can't use it to explain all observations.

The atom was first theorized by the Greek philosopher Democritus. Even in his day the discussion of matter was disputed. Democritus and Aristotle argued over the description of matter. It is interesting to think of these two people discussing something they couldn't see. We respect their arguments now, but their discussions might not have been much different than the ones you might have with your friends. One significant difference would be that they had their discussion in public, and the result of their discussion was eventually written down. Science advances today through public debate and publication. Not much different really.

It took almost 2000 years before Democritus' ideas came back around. Dalton and Lavoisier separately studied matter and chemical reactions in the late 1700s. They proposed rules and definitions, tested these ideas, discussed them, and finally an updated view of the atom was proposed.

Still, people were asking, "What would an atom look like if we could see it with our eyes?" A research group led by J.J. Thomson started doing experiments to probe the atom further. Thomson, E. Goldstein, and Sir James Chadwick performed experiments for more than 35 years. The result of their work is the current "view" of the atom. In this chapter, we will probe their view in order to "see" an atom.

Chapter Goals

- Define the three major particles in the atom

- Understand the relative distance between the nucleus and the electrons

- Learn the charge on each particle

- Be able to estimate the relative mass of each particle

- Understand which subatomic particles are added or removed when an ion is formed from an atom

- Recognize how our chemical symbols communicate the number of subatomic particles that have been added or removed

- Understand how mass and atomic numbers are represented with a chemical symbol

ACTIVITY 5.1

Objective

• Define the three major particles in the atom

Getting Started

We represent atoms as spheres. Although this representation is useful most of the time, it does not explain fundamental differences in the ways elements interact with light or with each other. J.J. Thomson was the first scientist to discover a subatomic particle. Ten years later Thomson and Goldstein found a second subatomic particle, and about 20 years after that Chadwick found a third subatomic particle. Study the Model to gain the perspective of the atom that resulted from their experiments and discussion.

The Model

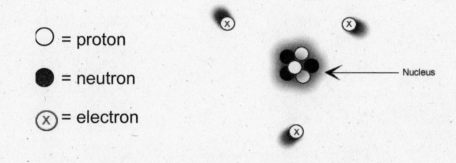

Figure 5.1   A Model of The Atom (Not to Scale)

Reviewing the Model

1.   How are the neutrons represented in the above model? _____

2.   How many electrons are shown in the above model? _____

Exploring the Model:

3.   The nucleus is the term for the center of the atom, as shown in Figure 5.1. Which particles are present in the nucleus of an atom?

4.   The nucleus is referred to as a dense part of the atom that contains most of the mass of an atom. What does this tell you about the masses of protons and neutrons as compared to electrons?

Exercising Your Knowledge

5.   In examining the Model, which particle would be expected to be easiest to remove from an atom and why?

Summarizing Your Thoughts

6.   Have your instructor review your answers. To receive all possible points you must have the instructor's initials here: _____

ACTIVITY 5.2

Objective

- Understand the relative distance between the nucleus and the electrons

Getting Started

As we noted in the last model, Figure 5.1 is not drawn to scale. The idea of the scale of the atom and where the particles are located puzzled many scientists for a long time until Ernest Rutherford (along with his students Hans Geiger and Ernest Marsden) ran his famous gold-foil experiment. Our model here is an oversimplification of Rutherford's experiment, but we can use it to gain some sense of the size of the nucleus with respect to the atom.

The Model

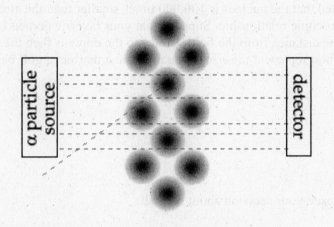

Figure 5.2   Simplified Rutherford Experiment

- Only 1 in 20,000 α-particles are deflected.

- Spheres represent the atoms in the gold foil. We represent three layers of gold; Rutherford's foil would have had many more layers.

- (Alternate Model: lay ten coins on the table in an arrangement similar to the Model shown. Now think about how easy it should be to draw a line through the coins without disrupting their arrangement.)

Reviewing the Model

1. What percentage of the α-particles was deflected?

2. Rutherford estimated the diameter of the atom to be 10^{-8} cm. Estimate the maximum number of layers in his foil if the foil was 0.1 mm thick. (Hint: lay the spheres in a straight line with each edge touching.)

Exploring the Model

3.   If the spheres were solid, how many particles would reach the detector?

4.   Speculate about why any particle would be deflected back toward the source.

5.   In the Model presented in Activity 5.1, the electrons are shown outside the nucleus. On the basis of how many particles hit the detector in Activity 5.2, is it reasonable to think that the α-particles collide with the electrons?

Exercising Your Knowledge

6.   Rutherford estimated that the nucleus is 100,000 times smaller than the atom. Let's relate this estimate to a macroscopic relationship. Suppose that your favorite domed football stadium is one-half of an atom. The distance from the field surface to the dome is then the radius of the atom. Which ball would best represent the size of the nucleus: a marble, a golf ball, a soccer ball, or a beach ball?

7.   Explain how you made your decision about the ball.

ACTIVITY 5.3

Objectives

- Learn the charge on each particle
- Be able to estimate the relative mass of each particle

Getting Started:

Now that we have identified the parts that make up an atom, it will be important for us to examine how changing the number of each particle changes two important properties: the charge and the mass. Review the Model to determine the fundamental charge and mass of the three particles.

The Model:

◯ = proton   ● = neutron   (x) = electron

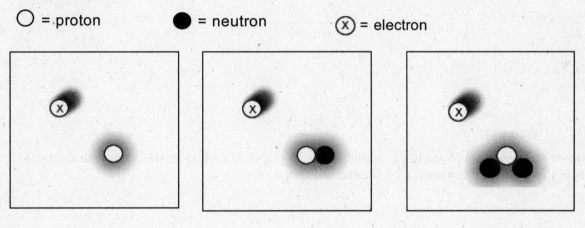

Atom A	**Atom B**	**Atom C**
Atomic mass ≈ 1 amu	2 amu	3 amu

- The unit used for measuring the mass of an atom is the atomic mass unit (amu).

- The proton has a positive charge (+1). The overall charge for all three atoms in the Model is zero.

Reviewing the Model:

1. In atom A how many electrons are there? _____ In B? _____ In C? _____

2. In atom A how many protons are there? _____ In B? _____ In C? _____

3. In atom A how many neutrons are there? _____ In B? _____ In C? _____

Exploring the Model:

4. If a proton has a mass of approximately 1 amu, estimate the mass of the electron. _____ amu

5. Estimate the mass of a neutron. _____ amu

6. What is the charge on the electron? _____

7. What is the charge on the neutron? _____

8. If it were possible to add another neutron to the atom depicted in the Model, what would be its approximate atomic mass? _____

Summarizing your Thoughts:

9. Prepare a table that will allow you to quickly remember the name, charge, approximate mass, and the location of each atomic particle.

10. Atoms A, B, and C depicted in this activity are all *isotopes* of hydrogen. Based on your thoughts about the Model, write a sentence to define an isotope.

11. For your entire group there are five cookies available for participation. Using only whole cookies, divide these five cookies among your group members. The distribution of cookies should indicate how much each individual contributed to the completion of this activity. Write the name and number of points for each group member below.

ACTIVITY 5.4

Objectives

- Understand which subatomic particles are added or removed when an ion is formed from an atom

- Recognize how our chemical symbols communicate the number of subatomic particles that have been added or removed

Getting Started

Though we will not get into all of it now, the addition or removal of one of the subatomic particles is important to the chemical reactions in which an atom can be involved. In later chapters we will look at electrons and how they affect reactions in much more detail. But for now, use the Model to learn which particles we are talking about, and how we symbolize the change in the number of these particles.

The Model

$\bigcirc$ = proton $\textcircled{x}$ = electron

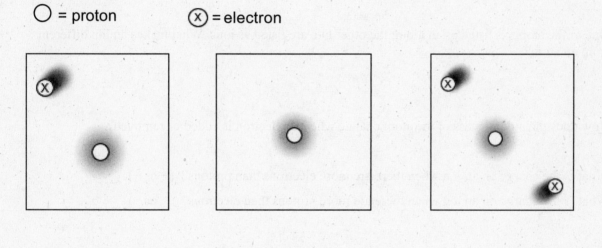

Atom A	Ion D	Ion E
Atomic mass 1 amu	1 amu	1 amu

$$D = {}^{1}_{1}H^{+} \longleftarrow \text{denotes +1 (positive charge)}$$

$$E = {}^{1}_{1}H^{-} \longleftarrow \text{denotes -1 (negative charge)}$$

Reviewing the Model

1. How many protons are in atom A? _____   In ion D? _____   In ion E? _____

2.   How many electrons are in atom A? _____   In ion D? _____   In ion E? _____

3.   How do the two chemical symbols differ?

Exploring the Model

4.   How are atom A, ion D, and ion E the same?

5.   How are they different?

6.   One of the atoms is listed as an atom; the other two are listed as ions. What makes an ion different from an atom?

7.   How does the atomic mass of the atom change when an electron is added or removed?

8.   What is the charge on an ion when there are more electrons than protons? (+ or −) _____

9.   What is the charge on an ion when there are more protons than electrons? (+ or −) _____

10.   Complete this definition: A _____ has more protons than electrons.

11.   Complete this definition: A _____ has more electrons than protons.

12.   What does this difference tell about the atoms that these chemical symbols represent?

Exercising your Knowledge

13.　Classify each row in Table 5.1 as a cation, anion, or atom.

Table 5.1

Number of Neutrons	Number of Electrons	Number of Protons	Classification
9	10	9	Anion
13	12	12	
32	28	30	
12	10	11	
5	4	4	
20	18	19	
16	18	16	
33	28	29	

14.　Have your instructor review your answers before moving on to the next section. Instructor's initials: _____

Summarizing Your Thoughts

15.　Summarize what distinguishes an atom from an ion.

16.　It sometimes surprises students that we use a − symbol to represent the ion where an electron has been *added*. Write yourself a note that will remind yourself when to use the + and − signs for ions.

17.　Compared to the last group session how well did your group work together?

18.　How well did your group stay on task through the course of the exercise?

ACTIVITY 5.5

Objective

- Understand how mass and atomic numbers are represented with a chemical symbol

Getting Started:

We have seen throughout this book that symbols make up a large part of chemistry because without them it would be difficult to communicate all the information we gather experimentally. We will now continue with our study of the atom by taking a look at how we can represent symbolically all the information we have obtained about the number of each particle in the atom.

The Model

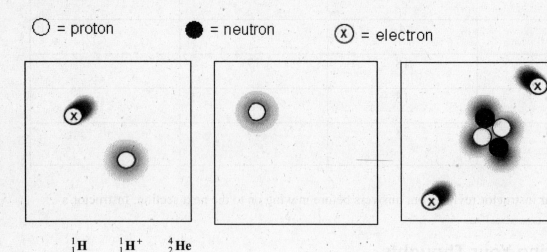

$\bigcirc$ = proton   ● = neutron   ⊗ = electron

$_{1}^{1}\text{H}$ $_{1}^{1}\text{H}^{+}$ $_{2}^{4}\text{He}$

Reviewing the Model

1. What elements are represented in the Model?

2. Which symbol represents the number of protons? (choose one)
    a.   the superscript before the letter
    b.   the subscript before the letter only
    c.   the subscript before the letter and the letter

Exploring the Model:

3. Describe the differences between the hydrogen atom and the helium atom.

4. Does the atom $_{2}^{2}\text{H}$ exist? How do you know this?

5. What would be the chemical symbol (with atomic and mass numbers) for the atom presented in figure 5.1? Check this answer with your instructor. Instructor Initials: _____

Summarizing Your Thoughts

6. A generic symbol for an element is shown. Write a key to this symbol that defines A, E, and Z.

$^{A}_{Z}\text{E}$

7. The mass number for the helium atom is four. Review all the atoms shown in this activity, and write a definition for mass number.

8. What information around a chemical symbol changes with a changing number of neutrons?

9. What information around a chemical symbol changes with a changing number of electrons?

10. Given two chemical symbols with atomic number, mass number, and charge, how could you tell if the two particles represented were isotopes?

CHAPTER 6

Significant Figures Are Not Just for History Class (Scientific Notation and Significant Figures)

ACTIVITY 6.1

Objective

- Recognize how a reported value is related to the method of measurement.

Getting Started

When scientists look at a measured value, they will consider the value's accuracy and precision. This thought process takes into account how the measurement was made and whether or not the measurement could be improved. For us all to understand each other, there must be some set of rules we can all understand. Following these rules is an ethical responsibility, so it is important to learn how we communicate. Review this initial model to see how the reported value relates how a value was measured.

The Model

Figure 6.1

Two students were asked to determine the number of beans in the jar shown in Figure 6.1. Each student took the jar home and came back to report.

- Anni: There are 484 beans in the jar
- Bart: There are 500 beans in the jar.

Reviewing the Model

1. What is the difference between the number of beans Anni reported and the number of beans Bart reported?

2. Is it possible to have two different values for the number of beans in the jar?

Exploring the Model

One of the students removed all the beans and counted them one at a time; the other one estimated the volume of a single bean to be about 1 cm^3, noted that the jar is labeled as 500 cm^3, and made an estimate. In the blanks provided, write the name of the student that adopted each procedure.

3.   Counted the beans: _____

4.   Estimated on the basis of size: _____

5.   Describe your reasons for making your assignments.

Exercising Your Knowledge

6.   How can a reported number help you determine the most accurate measurement of a quantity?

7.   Circle the value in each pair that was measured using the most accurate method.

   a.   10,000 people or 9,468 people attended the game

   b.   1 million or 1.0 million dollars

8.   Suppose you were given a million dollars in pennies. Which method would provide the most accurate count if you had only 2 hours to determine whether or not you had been cheated: counting the pennies or weighing all the pennies and dividing by the average mass of a single penny? Explain your choice.

Summarizing Your Thoughts

9.   Using the example of the million pennies, explain why it is sometimes necessary to use methods that don't give an exact answer.

10.   A significant figure is related to how well you know a number. Anni's value is considered to possess three significant figures. Bart's value possesses only one significant figure. In the preceding sentences what does *figure* mean?

11. Review your definition of *figure* and the values from the Model, and explain what is meant by "significant."

12. It is your responsibility to report the number of spiders in the forest. You decide that it is not possible to count them all, so you use a method that requires some approximations. Would it be ethical to report a value such as 1,632,388 spiders if you didn't count all of them? Why?

ACTIVITY 6.2

Objectives

- Recognize the conventions of scientific notation
- Gain a sense of the magnitude of a value reported in scientific notation

Getting Started

We deal with very large and very small numbers throughout chemistry. It is often difficult to work with values of these magnitudes, especially when we may know the value to only one or two significant figures. Thus we employ a notation that helps us write these values in a simpler manner. However, we need to make sure we can quickly grasp a value's magnitude when we look at it. This activity introduces the notation we use and explores how this notation should give us a sense of a value's magnitude.

The Model

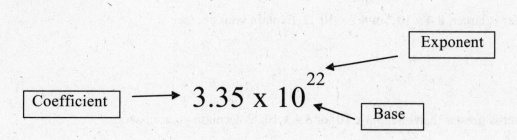

Table 6.1

Value	In Scientific Notation
3097	3.097×10^3
616	6.16×10^2
-59.67	-5.967×10^1
0.000138	1.38×10^{-4}
0.12	1.2×10^{-1}

Reviewing the Model

1. Which values are greater than one?

2. Which values are less than one?

3.   Which value is a negative number?

4.   Represent the value 616 in scientific notation: _____

5.   Represent the value 0.12 in scientific notation: _____

6.   In the value 6.16×10^2, which number is the exponent?

Exploring the Model

7.   Which number is larger, 2.3×10^3 or 8.4×10^2? Explain your choice:

8.   Which number is larger, 8.4×10^{-4} or 4.7×10^{-1}? Explain your choice.

9.   Which number is greater than one: 3.3×10^3 or 3.4×10^{-4}? Explain your answer.

Exercising Your Knowledge

10.   Label the number line, and place an "X" where you expect to find the value 5.2×10^3.

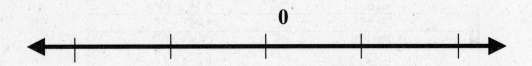

11.   Label the number line, and place an "X" where you expect to find the value 5.2×10^{-3}.

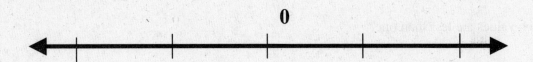

12. Label the number line, and place an "X" where you expect to find the value -5.2×10^3.

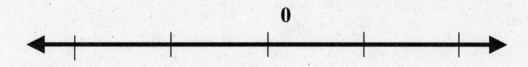

0

13. Complete the table:

Table 6.2

Value	In Scientific Notation
583	
	8.246×10^2
−3291	
0.265	
	4.37×10^{-3}
	7.28×10^1
−0.00000821	

Summarizing Your Thoughts

14. Why are some exponents positive and others negative?

15. Consider the two values, 3.67×10^{-3} and -3.67×10^3. What is the difference in meaning of the negative symbol (−) in these two values?

16. When writing a number in scientific notation, where do you place the decimal point?

17. Explain which parts of a value written in scientific notation should be considered and the order in which they are compared when you compare two values written in scientific notation to determine their relative magnitude (largest or smallest).

ACTIVITY 6.3

Objective

- Understand how zeros are used when writing values in scientific notation

Getting Started

Zeros cause a great deal of confusion when writing values in scientific notation. Sometimes they are significant, and other times they are simply indicating the magnitude of the number. Explore the Model to write a set of rules that helps you remember how to report zeros in a value.

The Model

Table 6.3

Value	Scientific Notation	Significant Figures
3097	3.097×10^3	4
−6001	-6.001×10^3	4
59.67	5.967×10^1	4
0.000128	1.28×10^{-4}	3
0.10	1.0×10^{-1}	2
−0.00504	-5.04×10^{-3}	3
600	6×10^2	1
600.	6.00×10^2	3

Reviewing the Model

1. Which values have zeros between two other numbers?

2. Which values have zeros at the beginning of the value?

3. Which values have zeros at the end of the value?

4. There are two values shown for six hundred; how are they represented differently?

Exploring the Model

Use the Model to help answer each question with a yes, no, or sometimes. If you decide to write "sometimes," provide an explanation of cases where you would answer "yes" or you would answer "no."

5. When a zero is between two numbers, is the zero included in the scientific notation?

6. When a zero is between two numbers, is it counted as a significant figure?

7. When a zero is at the beginning of a value, is the zero included in the scientific notation?

8. When a zero is at the beginning of a value, is it counted as a significant figure?

9. When a zero is at the end of a value, is the zero included in the scientific notation?

10. When a zero is at the end of a value, is it counted as a significant figure?

11. What is the relationship between the coefficient of a number in scientific notation and the number of significant figures?

12. Does it make a difference whether a value is positive or negative when counting the number of significant figures?

Exercising Your Knowledge

13. Complete the following table:

Table 6.4

Value	Scientific Notation	Significant Figures
2014		4
	-7.20×10^5	
0.0094		
5964.20		
	8.2×10^3	
	2.841×10^{-4}	

14. How can significant figures and scientific notation quickly help a reader assess how well we know a measured quantity?

15. If Anni predicted there were 4.82×10^2 beans in a jar, and Bart predicted that there were 5×10^2 beans, explain which of the two reported values is more precise.

    Have your facilitator review your work and initial here: _____

Summarizing Your Thoughts

16. What are some advantages of using scientific notation?

17. Work with your group to prepare a set of rules you could use later to convert a number into scientific notation.

ACTIVITY 6.4

Significant Figures (Addition and Subtraction)

Objective

- Write a set of rules for working with significant figures in addition and subtraction calculations

Getting Started

As you might suspect, as we perform calculations in chemistry we have to determine how to communicate the accuracy of a measurement that was made once the calculation has been completed. Estimations that are added to more accurate measurements need to be communicated. At the end of this exercise, you should be able to develop a rule for how to report a value to the proper number of significant figures after performing addition or subtraction on a set of values.

The Model

	A	B	C	D
	2.436	356.934	88.30	243.872
	+ 15.3	+ 5.11	+ 294.837	- 241
Calculator Shows:	17.736	362.044	383.137	2.872
Should Be Reported As:	**17.7**	**362.04**	**383.14**	**3**

Reviewing the Model

1. Write each reported value in scientific notation.

2. Which number in the addition portion of A has the fewest figures after the decimal point?

3. Which number in the answer portion of A (the "**Calculator Shows**" or **Should Be Reported As**") contains the fewest figures?

4. How many figures are located after the decimal in the "**Should Be Reported As**" value of B?

5.   How many figures are located after the decimal in each of the numbers being added in the B section?

Exploring the Model

6.   Is the reported value always one figure less than the value the calculator presents?

7.   Is the reported value always the same as the value with the fewest number of significant figures?

8.   How does the number reported by the calculator relate to the numbers that are initially added?

9.   Review the calculations and the reported value. Is the reported value related to the number of significant figures?

Summarizing Your Thoughts

10.   What trend or rule can help you determine the number of significant figures in added or subtracted answers?

ACTIVITY 6.5

Scientific Notation and Significant Figures (Multiplication and Division)

The Model

	A	B	C	D
	2.436	356.934	88.30	243.872
	x 15.3	x 5.11	÷ 294.837	÷ 241
Calculator Shows:	37.2708	1823.93274	0.299488	1.011917012
Should Be Reported As:	37.3	1820	0.2995	1.01

Reviewing the Model

1. What number in the multiplication portion of A has the fewest figures? How many are present in that number?

2. How many figures are located in the "**Should Be Reported As**" answer of A?

Exploring the Model

3. Which numbers contain fewer figures, the "**Calculator Shows**" or "**Should Be Reported As**" values?

4. Which value in the multiplication operators of B contains the fewest number of significant figures? How many?

5. How many significant figures are present in the "**Should Be Reported As**" answer of B?

6. Determine the number of significant figures in the "**Should Be Reported As**" answer of C, and compare it with the number of significant figures in the operators of division. What do you find?

Exercising Your Knowledge

7. Perform the following operations, and report your answer to the proper number of significant figures.

    a.   15.3 x 44.365 =

    b.   6.44 ÷ 57 =

Summarizing Your Thoughts

8. What trend or rule can help you determine the number of significant figures in multiplied or divided answers?

9. Why does the "**Should Be Reported As**" row contain fewer significant figures than the "**Calculator Shows**" values?

ACTIVITY 6.6

Using Scientific Notation in Calculations

Objectives

- Understand the mathematical rules that define scientific notation operations
- Perform operations using numbers in their scientific notation

Getting Started

When large numbers are added, subtracted, multiplied, and divided, a large number of zeros or additional coefficients make these processes challenging. Using scientific notation can simplify these steps.

The Model

Table 6.5

Mathematical Operation	Normal Notation	Scientific Notation	Answer
Addition	$5,000 + 1,000 =$	$5.0 \times 10^3 + 1.0 \times 10^3 =$	6.0×10^3
Subtraction	$5,000 - 1,000 =$	$5.0 \times 10^3 - 1.0 \times 10^3 =$	4.0×10^3
Multiplication	$5,000 \times 1,000 =$	$5.0 \times 10^3 \times 1.0 \times 10^3 =$	5.0×10^6
Division	$5,000 / 1,000 =$	$5.0 \times 10^3 / 1.0 \times 10^3 =$	5.0×10^0

Reviewing the Model

1. What is 5,000 in scientific notation?

2. According to Table 6.5, what is the answer for $5,000 + 1,000$ in scientific notation?

3. What is the answer for $5,000 \times 1,000$ in scientific notation?

4. In the value 5.0×10^0, what does it mean to multiply by 10^0?

Exploring the Model

5. Does the exponent in scientific notation change if it is added to a number containing the same scientific notation exponent? What if these numbers are subtracted?

6. What happens to the coefficient when it is added in scientific notation? What about subtracted?

7.   Refer to the Model, and explain what happens to exponents if numbers in scientific notation are multiplied.

8.   What happens to coefficients when they are multiplied in scientific notation? What if they are divided?

9.   Refer to the Model, and explain what happens to exponents if numbers in scientific notation are divided.

Have your facilitator review your answers and initial here: _____

Exercising Your Knowledge

10.   Fill in the empty spaces in Table 6.6:

Table 6.6

Mathematical Operation	Normal Notation	Scientific Notation	Answer
	$200 - 0.3 =$	$2.0 \times 10^2 - 3.0 \times 10^{-1} =$	
	$63 + 500 =$		5.63×10^2
Multiplication		$6.0 \times 10^5 \times 1.0 \times 10^{-3} =$	
	$7,200 / 0.001 =$	$7.2 \times 10^3 / 1.0 \times 10^{-3} =$	

11.   If Anni weighed 135 pounds and Donnie weighed 175 pounds, how many total pounds would they weigh together? Show your calculation in scientific notation.

Summarizing Your Thoughts

12.   Describe how you can quickly estimate the magnitude of a value by using your observations about how the exponents change for the different mathematical operations.

CHAPTER 7

Periodic Advances at the Table (The Periodic Table)

Chapter Objectives

- Gain some sense of how the elements are organized on the periodic table
- Use this organization to make some simple predictions

ACTIVITY 7.1

Objective

- Organize a group of elements by some commonality

Getting Started

The story of how the modern periodic table was organized is quite amazing. At the time there were very few elements known, and everyone was trying to figure out some way to organize them in such a manner that you could predict the types of compounds and chemical reactions that would occur without placing yourself in danger. A very private Russian chemist named Dmitri Mendeleev proposed a simple, rational organization. Others at the time were offering a similar model, but Mendeleev made a few (at the time) outlandish predictions that came true. It is for the simplicity of his model and its ability to make good predictions that Mendeleev is most often credited with the modern periodic table. As we begin here, we won't reproduce Mendeleev's insight but rather try to see how we might start to group different elements together. Look at the Model to see how we might begin to organize the elements on the periodic table.

The Model

A simple anion: chloride, Cl^-

Some compounds containing chloride: $NaCl$, $CaCl_2$, $LiCl$, $BaCl_2$, $MgCl_2$, KCl

A simple cation: sodium cation, Na^+

Some compounds containing the sodium cation: $NaCl$, Na_2O, $NaBr$, NaN_3, NaF, Na_2S

Reviewing the Model

1. If the charge on the chloride ion is -1, then what is the charge on the lithium cation, LiCl? _____

2. If the charge on the sodium ion is $+1$, then what is the charge on the oxygen in Na_2O? _____

Exploring the Model

3.   List all the elements (ions) from the Model separately, and group them according to their charge.

4.   Circle the elements (ions) in your list on the periodic table.

1 H Hydrogen																	2 He Helium
3 Li Lithium	4 Be Beryllium											5 B Boron	6 C Carbon	7 N Nitrogen	8 O Oxygen	9 F Fluorine	10 Ne Neon
11 Na Sodium	12 Mg Magnesium											13 Al Aluminium	14 Si Silicon	15 P Phosphorus	16 S Sulfur	17 Cl Chlorine	18 Ar Argon
19 K Potassium	20 Ca Calcium	21 Sc Scandium	22 Ti Titanium	23 V Vanadium	24 Cr Chromium	25 Mn Manganese	26 Fe Iron	27 Co Cobalt	28 Ni Nickel	29 Cu Copper	30 Zn Zinc	31 Ga Gallium	32 Ge Germanium	33 As Arsenic	34 Se Selenium	35 Br Bromine	36 Kr Krypton
37 Rb Rubidium	38 Sr Strontium	39 Y Yttrium	40 Zr Zirconium	41 Nb Niobium	42 Mo Molybdenum	43 Tc Technetium	44 Ru Ruthenium	45 Rh Rhodium	46 Pd Palladium	47 Ag Silver	48 Cd Cadmium	49 In Indium	50 Sn Tin	51 Sb Antimony	52 Te Tellurium	53 I Iodine	54 Xe Xenon
55 Cs Cesium	56 Ba Barium	57 La* Lanthanum	72 Hf Hafnium	73 Ta Tantalum	74 W Tungsten	75 Re Rhenium	76 Os Osmium	77 Ir Iridium	78 Pt Platinum	79 Au Gold	80 Hg Mercury	81 Tl Thallium	82 Pb Lead	83 Bi Bismuth	84 Po Polonium	85 At Astatine	86 Rn Radon
87 Fr Francium	88 Ra Radium	89 Ac** Actinium	104 Rf Rutherfordium	105 Db Dubnium	106 Sg Seaborgium	107 Bh Bohrium	108 Hs Hassium	109 Mt Meitnerium									

Figure 7.1   The Periodic Table, Main Group and Transition Elements

5.   Does the table organize these elements according to your grouping based on charge?

6.   Does the order of the group within the column make any sense? What property is used to organize the elements within the column?

7.   Does there appear to be any relationship between an element's group (column) position and its charge?

Exercising Your Knowledge

8.   Fluorine is found in nature as an ion; what charge is expected for the ion of fluorine? _____

9.   What charge would you expect for the ion formed by strontium? _____

10. Predict the compound that would be expected from the common ions of calcium and oxygen.

11. Predict the compound that would be expected from the common ions of cesium and bromine.

Summarizing Your Thoughts

12. Explain how the organization of the elements in the periodic table can be helpful when predicting the charge on an ion and common compound formulae.

13. Write yourself a rule that helps you remember how to predict the common charges for the groups investigated in this activity.

ACTIVITY 7.2

Objective

- Relate the number of electrons available for bonding to the common charges

Getting Started

Interestingly, if you look back at our last activity, you will see that we avoided the groups in the middle of the periodic table (scandium to zinc). We did this for a reason: predictions for those elements are a bit more complicated (and we'll leave that for a general chemistry class). Thus, let us introduce a little bit of nomenclature: main group elements and transition elements. We will be working with the main group elements (those shown in Figure 7.2) and will leave the transition elements for another discussion in some other setting.

Also notice from the last model that we grouped the elements by their common charges. That should be a clue that the electrons are important. Recall from earlier activities that cations and anions result from the loss or gain of an electron, respectively. Thus, it is time to think a little farther into how the electrons could be important to the organization of the periodic table.

But before we begin we need a little thought game. Electrons are a big part of why atoms hang out together. If that is true and there are many electrons around, then we need to ask ourselves, "Are some electrons more important than others when compounds form?" The answer is a definite "yes"!

As we group the atoms, we will group them according to the electrons that are farthest away from the nucleus because they will be the electrons available to do chemistry. The electrons doing the chemistry are called *valence electrons*. There is a quantum mechanical description explaining which electrons we should consider (think messy calculus), but a simple view is to think about the periodic nature of things.

The idea of valence was discussed extensively and the use of dots in the Model was initially described by G.N. Lewis (not the esteemed co-author of this book). So the symbolism in the Model is often referred to as Lewis dot structures.

Use this activity to find the simple view and hope that as you progress in your studies you will add to your knowledge to gain a greater understanding.

The Model

Group 1A	Group 2A	Group 3A	Group 4A	Group 5A	Group 6A	Group 7A	Group 8A
H•							He:
Li•	•Be•	•B•	•C•	:N•	:O•	:F:	:Ne:
Na•	•Mg•	•Al•	•Si•	:P•	:S•	:Cl:	:Ar:
K•	•Ca•	•Ga•	•Ge•	:As•	:Se•	:Br•	:Kr:

Figure 7.2   **A Periodic Table Showing Only the Main Group Elements**

The common cation of lithium is Li$^+$. The lithium cation has two electrons, the same number of electrons that the neutral (no charge) helium atom has.

The common ion of fluorine is F$^-$. The fluoride ion has the same number of electrons as the neutral (no charge) neon atom.

Reviewing the Model

1.   If a fluorine atom contains nine electrons, how many electrons does the fluoride ion possess? _____

2.   How many total electrons are in a neon atom? _____

Exploring the Model

3.   Describe the properties of the Group 8A elements. What do you know about these elements? Have you ever heard of chemical compounds of these elements?

4.   For each of the Group 1A (a group is a column) elements, what is the expected charge? _____

5.   For each of the Group 7A elements, what is the expected charge? _____

6.   How many total electrons does potassium have when it forms its cation? _____

7.   How many total electrons does chlorine have when it forms its anion? _____

8.   How many total electrons does argon have? _____

9.   How many dots are drawn around argon? _____

10.   Let's let the number of dots be the available electrons. How do your observations about the ions of potassium and chlorine relate?

Exercising Your Knowledge

11.   For each dot structure, provide an example element, and then predict the common charge on that element:

:Ẍ: Element: _____   common ion of the element: _____

·Ẋ·   Element: _____   common ion of the element: _____

·Ẍ·   Element: _____   common ion of the element: _____

:Ẍ·   Element: _____   common ion of the element: _____

Summarizing Your Thoughts

12. Describe the relationship between the common ions and the number of electrons available to do chemistry (the number of dots we draw).

13. Write a couple of short sentences that will help you remember how the column position helps you remember how many dots to add or remove to determine the common ions.

14. Describe how your group completed this activity. Who led the discussion? Who recorded the answers? Did anyone disagree with the answers submitted?

CHAPTER 8

A Compound By Any Other Name (Classification of Ionic and Covalent Compounds and Their Nomenclature)

ACTIVITY 8.1

Objective

- Based on the elements present in a chemical formula, classify the compound as ionic or molecular.

Getting Started

Review the definitions for an element and a compound.

The Model

Fe is the elemental form of iron.

C is the elemental form of carbon.

Cl_2 is the elemental form of chlorine.

$FeCl_3$ is a compound formed from the elements iron and chlorine.

CCl_4 is a compound formed from the elements carbon and chlorine.

Reviewing the Model

1. What does the subscript 2 indicate in Cl_2?

2. What is implied when there is no subscript?

3. Classify C, Fe and Cl as metals or nonmetals.

Exploring the Model

4. Evaluate the statement, "The formulas for elements never contain a subscript." Is this statement true?

5. From the two examples provided, would you expect the formula S_8 to represent a compound or an element?

6.   Using examples from the Model, explain how you classified S_8.

Exercising Your Knowledge

7.   Classify each formula below as an element or a compound.

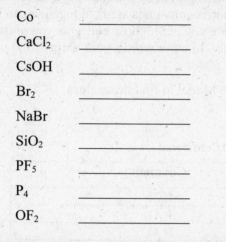

Co _____

$CaCl_2$ _____

CsOH _____

Br_2 _____

NaBr _____

SiO_2 _____

PF_5 _____

P_4 _____

OF_2 _____

Summarizing Your Thoughts

8.   What clues are given in a chemical formula that allow you to differentiate between an element and a compound?

ACTIVITY 8.2

Getting Started

There are two major classes of compounds typically encountered as part of an introductory course: ionic and covalent compounds. The concepts describing how these compounds are held together can be developed as you progress through your studies. However, before you get to those concepts you must be able to quickly classify a compound into one class or the other. In other words, your ability to classify compounds will guide how you will think about bigger ideas.

There are clues in the chemical formula. It is your job to use the Model to find these clues.

The Model

Table 8.1   Compounds That Are Considered...

Ionic	Covalent
$ZnCl_2$	CCl_4
Na_2O	P_2O_5
Fe_2O_3	N_2O_4
CuI	NI_3

Reviewing the Model

1.   The compound $ZnCl_2$ is considered to be a (an) _____ compound.

2.   The compound that contains nitrogen and oxygen is a (an) _____ compound.

3.   List all the elements included in the Model.

Exploring the Model

4.   Does the classification seem to be made based on how many atoms of each element are represented in the formula?

5.   Write the symbols for the elements presented by the Model near their correct location on the outline of the periodic table.

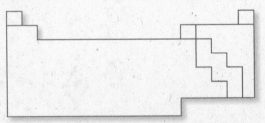

Figure 8.1   Outline of The Periodic Table, Main Group and Transition Elements

6. Compare the types of elements found (metals or nonmetals) for the two classes of compounds. Do you see any trend in the type of elements present and the classification?

Exercising Your Knowledge

7. Classify each of the following as either ionic or covalent.

NaBr _____

SF_6 _____

$CoBr_2$ _____

OF_2 _____

NO_2 _____

BaS _____

CsF_2 _____

$CrCl_3$ _____

CO_2 _____

CO _____

Summarizing Your Thoughts

8. Write a simple rule that will allow you to classify compounds as ionic or covalent on the basis of what you have learned from the Model.

ACTIVITY 8.3

Objective

- Identify some simple rules about nomenclature (naming)

The Model

Examine the table below, and answer the following questions.

Table 8.2

Cation	Anion	Chemical Formula	Compound Name
Na^+	Cl^-	NaCl	sodium chloride
Ca^{2+}	O^{2-}	CaO	calcium oxide
Zn^{2+}	Cl^-	$ZnCl_2$	zinc chloride
Li^+	S^{2-}	Li_2S	lithium sulfide
K^+	N^{3-}	K_3N	potassium nitride

Reviewing the Model

1. Are ALL cations positive ions or negative ions? _____

2. Are ALL anions positive ions or negative ions? _____

3. What is the name of the compound formed by the combination of Li^+ ions and S^{2-} ions?

Exploring the Model

4. When the name of an ionic compound is given, which ion is stated first?

5. Compare the first part of the compound names to the name of the element from the periodic table. How does the name of the cation correspond to the name of the element?

6. Compare the second part of the compound name to the name of the element from the periodic table. How does the name of the anion correspond to the name of the element?

7. From what part of the periodic table do the cations in the Model come (metals or nonmetals)?

8.   From what part of the periodic table do the anions in the Model come?

Exercising Your Knowledge

9.   For each of the following, predict whether the ion will likely be a cation or an anion.

      a.   magnesium ion
      b.   selenide ion
      c.   bromide ion
      d.   cesium ion

10.   For each ionic compound, identify the cation and the anion.

      a.   sodium fluoride
      b.   strontium sulfide
      c.   lithium iodide
      d.   barium chloride

11.   In what way did the name provide clues about the classification of each element as a cation or anion?

12.   Where on the periodic table would you expect to find elements that ionize to form cations?

13.   Where on the periodic table would you expect to find elements that ionize to form anions?

Summarizing Your Thoughts

14.   Consider the clues you identified, and write a general rule for how you change the name of elements to cations when naming ionic compounds.

15.   Consider the clues you identified, and write a general rule for how you change the name of elements to anions when naming ionic compounds.

16.   Given the chemical formula of an ionic compound, list at *least* three necessary steps to give the correct name of that compound. (If needed, use a chemical formula of a compound from the table above as an example in listing the naming steps)

ACTIVITY 8.4

Predicting the correct chemical formula for ionic compounds formed from simple anions

Objective

- To learn how to predict the correct number of cations or anions in a simple salt

The Model

Examine the table below, and answer the following questions

Table 8.3

Cation	Anion	Chemical Formula	Compound Name
Na^+	Cl^-	$NaCl$	sodium chloride
Zn^{2+}	Cl^-	$ZnCl_2$	zinc chloride
Na^+	S^{2-}	Na_2S	sodium sulfide
K^+	N^{3-}	K_3N	potassium nitride

Reviewing the Model

1.   What is the charge on the zinc ion?   _____

2.   What is the charge on the nitride ion?   _____

3.   What is the charge on the chloride ion?   _____

4.   What is the charge on the ionic compound, sodium chloride?   _____

5.   What is the charge on the ionic compound, sodium sulfide?   _____

6.   How many potassium ions are present in K_3N?   _____

7.   What does the '2' stand for in the formula for $ZnCl_2$?   _____

Exploring the Model

8.   Sodium chloride is $NaCl$, and zinc chloride is $ZnCl_2$. Why are there more chloride ions in the zinc compound?

9.   Sodium chloride is $NaCl$, and sodium sulfide is Na_2S. Why are there more sodium ions in the sulfide compound?

Exercising Your Knowledge

10.   How many chloride ions would combine with an Al^{3+} ion to form aluminum chloride?

11.   What charge does the barium ion possess in the compound $BaCl_2$?

Summarizing Your Thoughts

12.   Explain how you determined the number of chloride ions needed in aluminum chloride.

13.   From the table and the answers above, what do you know about the overall charge on ALL ionic compounds?

14.   List at *least* three necessary steps to obtain the correct formula of any simple ionic compound.

15.   List all the members of your group in the order in which they contributed to the successful completion of this activity.

16.   What can you do to make sure that all group members understand the material presented?

Have the instructor check your answers above, and initial here: _____

ACTIVITY 8.5

Objective

• Understand how to write the chemical formula of ionic compounds containing metals with varying oxidation states

Getting Started

When a Group 1A metal forms a cation, it will always form a +1 cation. When a Group 2A metal forms a cation, it will always form a +2 cation. However, as we progress into the transition metals we find that these metals can form cations with different charges under different circumstances. Use the Model below to develop some rules that describe how to communicate the charge of the cation.

The Model

Examine the table below, and answer the following questions

Table 8.4

Chemical Formula	Compound Name
$FeBr_2$	iron(II) bromide
$FeBr_3$	iron(III) bromide
PbO	lead(II) oxide
PbO_2	lead(IV) oxide
Cu_3N	copper(I) nitride
Cu_3N_2	copper(II) nitride

Reviewing the Model

1. What is the expected charge on the bromide ion?

2. What is the expected charge on the oxide ion?

3. What is the expected charge on the nitride ion?

4. Represent the Roman numeral II as an Arabic numeral.   _____

5. Represent the Roman numeral III as an Arabic numeral.   _____

6. Represent the Roman numeral IV as an Arabic numeral.   _____

7. What two elements are present in the compounds in the last two rows of the table?

8. What is different about the *chemical formulas* of these last two compounds?

9. What is different about the *compound names* of these last two compounds?

Exploring the Model

10.   Use your rules developed in Activity 8.4 to determine the charge on the iron ion in these compounds:

        Charge on iron in $FeBr_2$:

        Charge on iron in $FeBr_3$:

11.   How is the Roman numeral in the compound name related to the charge on the iron atoms?

12.   Does this hold true for all the compounds in the table above?

13.   What types of metals require the use of a Roman numeral in the name of their ionic compounds?

14.   Where are these metals located on the periodic table?

Exercising Your Knowledge

15.   Why do the compounds in this activity require Roman numerals in the name while compounds such as calcium chloride do not?

16.   If only the chemical formulas were given for the compounds in the above examples, how could you determine the amount of charge on the cation?

17.   Complete the table that follows with the proper ions, chemical formulas, and compound names. The first row has been completed as an example.

Table 8.5

Cation	Anion	Chemical Formula	Compound Name
Na^+	Cl^-	NaCl	sodium chloride
Ba^{2+}	I^-	BaI_2	
Mn^{2+}	O^{2-}		manganese(II) oxide
Mg^{2+}	N^{3-}		
			cobalt(III) fluoride
		CrO	
Cu^+	S^{2-}		
		Ca_3P_2	
		SnS_2	

Summarizing Your Thoughts

18. How will you know when to use a Roman numeral when writing the name of an ionic compound?

19. Look at your answer from Activity 8.4 that lists the steps necessary to give the correct name of an ionic compound from its chemical formula. How do the steps differ when ions with varying oxidation states are involved?

20. Look at your answer from Activity 8.5 that lists the steps necessary to give the correct chemical formula of an ionic compound given its name. How do the steps differ when ions with varying charges are involved?

ACTIVITY 8.6

Objective:

- Recognize the names of polyatomic ions, and understand how to write the chemical formulas of ionic compounds containing polyatomic ions

Getting Started

Thus far we have considered only simple, monoatomic cations and anions. There is another class of ions that are often called polyatomic ions. Polyatomic ions are a group of atoms that are held together by covalent interactions, and the entire group of atoms carries the charge. The most common polyatomic ions contain oxygen. Their names may not seem to make sense now, but there is a system to this madness. It is your task to study the Model and determine what the nomenclature rules are.

The Model

Table 8.6

Ion	Name	Ion	Name
N^{3-}	nitride	S^{2-}	sulfide
NO_2^-	nitrite	SO_3^{2-}	sulfite
NO_3^-	nitrate	SO_4^{2-}	sulfate

Reviewing the Model

1. What element is associated with the prefix "nitr-"?

2. What element is associated with the prefix "sulf-"?

3. What is the ending (suffix) when there are no oxygen atoms in the formula?

4. What suffixes are used when oxygen is included in the formula?

Exloring the Model

5. Does the suffix of each name depend on the charge of the ion?

6. Does the suffix tell you how many oxygen atoms there are?

7. Compare nitrate to nitrite. Which ion has more oxygen atoms?

8. Compare sulfate to sulfite. Which ion has more oxygen atoms?

Exercising Your Knowledge

9. Consider the two oxo- ions of chlorine, ClO_2^- and ClO_3^-. Which ion would have the –ate ending?

    a. Write the names of these two oxo- ions of chlorine.

10.   Consider the two oxo- ions of phosphorus, PO_3^{2-} and PO_4^{3-}. Which ion would have an –ate ending?

     a.   Write the names of these two oxo- ions of phosphorus.

11.   In the movie *Star Wars Episode V: The Empire Strikes Back*, the character Han Solo is frozen in the fictional material, carbonite. If the carbonite ion existed, what would be its likely chemical formula?

12.   Why would it be unlikely for a solid material to be made entirely of pure carbonite ions?

Summarizing Your Thoughts

13.   The last three letters of a name can tell a lot about a particle! For each of the name endings below, give a general description of what type of ion or particle would be expected to have that ending (cation, monatomic anion, polyatomic anion, metal element, and/or nonmetal element).

     a.   –ide   _____

     b.   –ium   _____

     c.   –ate   _____

     d.   –ine   _____

     e.   –ite   _____

Have the instructor check your answers above, and initial here: _____

·ACTIVITY 8.7

Getting Started

Now that we have introduced polyatomic ions, we have to consider how this new twist affects the name of a compound and how we write the chemical formula. There are new features in the Model; let's see if we can figure out how to handle them.

The Model

Table 8.7

Chemical Formula	Compound Name
$CaSO_4$	calcium sulfate
$CaSO_3$	calcium sulfite
Na_3PO_4	sodium phosphate
Li_2CO_3	lithium carbonate
NH_4Cl	ammonium chloride
$Be(NO_2)_2$	beryllium nitrite
$Mg_3(PO_3)_2$	magnesium phosphite
$Fe(NO_3)_3$	iron(III) nitrate
$Al(OH)_3$	aluminum hydroxide

Reviewing the Model

1.   Write the name and symbol for all the monatomic ions in the Model.

2.   Write the name and formula (including the charge) for all the polyatomic ions in the Model.

3.   The ammonium cation is the only polyatomic cation in the Model. What are the formula and charge of the ammonium cation?

4.   How many nitrite ions are present in beryllium nitrite?   _____

5.   What new typesetting feature is used in these chemical formulas?

Exploring the Model

6.   Do all polyatomic ions require the use of parentheses?

7.   When are parentheses used?

8.   Have the nomenclature rules you established earlier in this chapter changed? If so, how? If not, is that important to know?

9.   How many nitrogen atoms are in beryllium nitrite?   _____

10. Describe the thought process you used to determine the number of nitrogen atoms in beryllium nitrite.

11. How many oxygen atoms are in beryllium nitrite?   _____

12. In what way does determining the number of oxygen atoms differ from the process you just described for nitrogen?

13. If the parentheses were omitted and aluminum hydroxide was written as $AlOH_3$, how would that change the number of atoms of each element present in the compound?

Exercising Your Knowledge

14. How many of each element is present in aluminum hydroxide?

    a.   Aluminum: _____

    b.   Oxygen: _____

    c.   Hydrogen: _____

15. Complete the table below with the proper ions, chemical formulas, and compound names. The first row is completed as an example.

Table 8.8

Cation	Anion	Chemical Formula	Compound Name
Na^+	Cl^-	NaCl	sodium chloride
		LiCN	lithium cyanide
Ca^{2+}	OH^-		
Fe^{2+}	NO_3^-		
			barium phosphate
Cr^{2+}	PO_3^{3-}		
K^+	SO_3^{3-}		
			ammonium carbonate
		$AuPO_3$	
			copper(II) cyanide

Summarizing Your Thoughts

16. Write a rule that can be used to determine whether or not parentheses are needed when writing a chemical formula.

17. Make a list of information you must know in order to write the correct formula for an ionic compound (this list may require reviewing all the activities in this chapter).

18. Explain why the number of each ion is not included in the name of an ionic compound.

ACTIVITY 8.8

Objectives

- Given a covalent compound's name, be able to give the proper chemical formula for the compound

- Given a covalent compound's chemical formula, be able to give the proper name for the compound

Getting Started

We will be using gases and other compounds as illustrations of naming covalent compounds. Covalent compounds are defined as groups of atoms that stay together because of shared electrons in chemical bonds. There are an infinite number of covalent compounds. Here, we will be focusing on naming some of the smaller covalent compounds.

The names of covalent compounds are similar to those of the ionic compounds, but there are differences. Use the Model to see if you can figure out how the rules differ.

The Model

Table 8.9

Compound Name	Compound Molecular Formula
phosphorus hexafluoride	PF_6
tetracarbon decahydride	C_4H_{10}
boron trichloride	BF_3
dinitrogen oxide	N_2O
carbon monoxide	CO
dinitrogen tetroxide	N_2O_4

Reviewing the Model

1. Where on the periodic table do you find all the elements used in the Model?

2. What suffix is used for all the compounds? _____

3. Is the name of the first element in each formula changed as it goes from an individual element to a compound? _____

4. How is the name of the second element in each name changed as it goes from an individual element to a compound?

5. How many atoms of nitrogen are present in dinitrogen tetroxide? _____

6. How many atoms of fluorine are present in phosphorus hexafluoride? _____

7. Use the Model to fill the following table with prefixes used to designate the number of each type of atom in a binary compound:

Table 8.10

Prefix	Number of atoms of element
(no prefix)	
	One
	Two
	Three
	Four
	Five
	Six
Hepta-	Seven
	Eight
	Nine
	Ten

Exploring the Model

8. Nitrogen oxide and nitrogen monoxide are both acceptable names for the compound with the chemical formula NO. What is different in these names, and why do you think the prefix is optional?

9. A dentist calls you up and needs to order more laughing gas for his dental clinic. You check in a chemistry reference and find that the chemical name for laughing gas is dinitrogen monoxide. You may order N_2O, NO, or NO_2. One is the correct compound and the other two are toxic gases. Which should be ordered to keep the patients happy and alive?

Exercising Your Knowledge

From the information given above, complete the following table.

Table 8.11

Compound Name	Compound Molecular Formula
sulfur difluoride	
	PCl_3
silicon dioxide	
	H_2S
carbon tetraiodide	
	$SiBr_2$
	P_4O_{10}
carbon dioxide	

Summarizing Your Thoughts

10. In the biological process called respiration, we inhale oxygen and exhale *carbon dioxide*. When fossil fuels are burned, a toxic gas that may be produced is *carbon monoxide*. Explain why you wouldn't use the name "carbon oxide" for these molecules.

11. In complete sentences state the rules for naming a covalent compound, given the compound's molecular formula.

12. Compare your rules for naming covalent compounds with the rules you established for writing the names of ionic compounds. Make sure your rules clearly help you decide when you use the prefixes indicating the number of atoms and when you use Roman numerals.

Chapter Summary

13.   What are three major ideas you learned in this chapter?

14.   What new information did you learn today that adds to previous knowledge that you had about a topic?

CHAPTER 9

Counting To a Trillion Trillions (The Mole Concept)

Chapter Goals:

- Write the relationship between the mass and the number of particles (atoms or molecules) as an equality

- Understand the magnitude of this relationship

- Use the equality and the conversion factors that arise to convert from mass to particles or particles to mass

ACTIVITY 9.1

Objectives

- Understand the relationship between the mass of an element and the number of particles (the mole)

- Write equalities, and use these equalities to create conversion factors

The Model

Beaker 1	**Beaker 2**	**Beaker 3**
55.8 g of iron	111.6 g of iron	112 g of cadmium
1 mole of iron	2 mole of iron	1 mole of cadmium
6.02×10^{23} atoms of iron	12.04×10^{23} atoms of iron	6.02×10^{23} atoms of cadmium

Reviewing the Model

1. Which beaker has more atoms of iron?

2. How many grams of iron in Beaker 1? In Beaker 2?

3. How many moles of iron in Beaker 1? In Beaker 2?

4. How many atoms of iron in Beaker 1? In Beaker 2?

Exploring the Model

5. Write the equality between grams of iron and moles of iron.

6. Write the equality between grams of cobalt and moles of cobalt.

7. Write the equality between moles of iron and atoms of iron.

8. Write the equality between moles of cadmium and atoms of cobalt.

9. Rearrange each of the above equalities (questions 5-8) into conversion factors.

10. Is the relationship between mass (grams) and number of particles the same for all three beakers in the Model?

11. Is the relationship between moles and atoms the same for all three beakers?

12. Compare the information in the Model for each element to the information provided on a periodic table. How is the mass (grams) related to the moles? How does the periodic table communicate this relationship?

Exercising Your Knowledge

13. Samples of different masses and numbers of particles for several elements are listed below. Complete the Table with the missing information for each element.

Table 9.1

Element	Mass of Sample	Number of Particles in Sample	Number of Moles in Sample
Magnesium		6.02×10^{23} atoms	1.00 mole
Arsenic	150 grams		
	23.0 grams	6.02×10^{23} atoms	
Lithium	13.9 grams		
	34.3 grams		0.25 moles
Boron		3.01×10^{23} atoms	
Silicon	56.2 grams		
	40.4 grams	12.04×10^{23} atoms	
Iodine			0.25 moles
	100 grams		0.50 moles

14. Complete the drawing by filling in the second beaker:

Beaker 1

55.8 g of iron

Beaker 2

77.2 g of iron

15. Explain how you made your estimate for the second beaker above.

16. If you have two samples, each with an actual amount of 100 g of silver and 100 g of gold, which sample has more atoms (or are they the same)? Explain your answer.

17. You have half a mole of nuts and three-quarters of a mole of bolts. Which sample has more pieces (or are they the same)? Explain how you arrived at the answer above.

Summarizing Your Thoughts

18. Based on your observations from the activity, explain how the number of atoms may be obtained from the number of moles of an element. Include in your explanation the information you might need to obtain from the periodic table.

19. Based on how you thought about which had more atoms, 100 g of silver or 100 g of gold, explain why it is difficult to compare chemical quantities using the mass of a sample.

20. Explain the benefit of thinking in terms of moles when working with chemical quantities.

21. In a complete sentence please describe the steps your group has taken to work through the problems more quickly.

22. Is everyone in your group contributing equally?

ACTIVITY 9.2

A Matter of Scale

The Model

$10^0 = 1$

$10^1 = 10$

$10^2 = 100$

$10^3 = 1\ 000$

$10^4 = 10\ 000$

$10^5 = 100\ 000$

10^1 is one power of 10 larger than 10^0

10^3 is two powers of 10 larger than 10^1

In other words, 1000 is 100 times larger than 10

The mole is defined as 6.02×10^{23} particles. Particles can be anything you can count as a single unit. Some examples:

- If you have a mole of apples, you have 6.02×10^{23} apples.

- If you have 6.02×10^{23} grains of sand on the beach, you have one mole.

- Given 6.02×10^{23} atoms of copper, you have one mole of copper.

Reviewing the Model

1.   How many more powers of 10 is 10^5 than 10^2?

2.   Restate the answer above another way ("In other words,…").

3.   Write the relationship between the mole and the number of particles as an equality.

4.   Write the two conversion factors that could be written from this equality.

Exploring the Model

5. Write all the zeros required to take 6.02×10^{23} out of scientific notation and write it as a simple number.

6. Compare the number of ^{1}H atoms needed for a mass of 1 gram with the number of average-sized paper clips needed to have a mass of 1 gram.

7. How many more powers of ten atoms are there in 1 gram of ^{1}H than there are seconds in 13.7 billion years? (Recall that 10^3 is two powers of 10 larger than 10^1.)

8. Is your answer for 13.7 billion years in seconds bigger or smaller than the number of 1 mole of seconds?

9. By how many powers of 10 does 13.7 billion years in seconds differ from 1 mole of seconds?

10. Could a chemist live long enough to individually count the number of gold atoms in a ring?

Exercising Your Knowledge

11. If you have 3.01×10^{23} apples, how many moles of apples do you have?

12. How many atoms do you have if you have 0.5 mole?

13. How many moles do you need to have 6,020,000 atoms?

Summarizing Your Thoughts

14. Explain in a few short sentences how to convert from moles of atoms to the number of atoms.

15. The magnitude of the mole is pretty big. Explain just how big this number is in language that could be understood by a small child.

CHAPTER 10

It's a Balancing Act

Objectives

- Understand what it means for a chemical reaction to be balanced

- Be able to balance a chemical equation

- Understand how mole ratios of reactants and products are communicated in a balanced chemical equation

- Be able to state and apply the Law of Conservation of Mass

ACTIVITY 10.1

Objectives

- Be able to balance a chemical equation

- Understand the Law of Conservation of Mass and how it applies to a balanced chemical equation

The Model

In studying for a chemistry exam, you become a bit hungry and decide to make a grilled cheese sandwich for a snack. Since chemistry is on your mind, you begin to think about this as a combination of bread and a slice of cheese, which may be represented as:

$$2\ Bd + Ch \rightarrow Bd_2Ch$$

Reviewing the Model

1.   What is the chemical symbol used to represent a slice of cheese?   _____

2.   What is the coefficient in front of the chemical symbol for bread?   _____

3.   What is the chemical formula representation for the product obtained?   _____

Exploring the Model

4.   Does the equation sufficiently describe the materials needed to make a sandwich?

5.   What does the coefficient represent in front of the chemical symbol for bread?

6.   How does a grilled cheese sandwich differ from two slices of bread and a slice of cheese?

7.   This sandwich only makes you hungrier, and you become inspired to make a larger sandwich of two slices of bread, a fried egg, two slices of ham, a slice of cheese, along with a Twinkie® (which come in packs of two), and three packets of mayonnaise. A chemical equation for this masterpiece is below. Put whole number coefficients in place to represent all the reactants being used up to make a whole number of sandwiches.

   ___ Bd + ___ Eg + ___ Hm + ___ Ch + ___ Tw_2 + ___ $My \rightarrow$ ___ $Bd_2EgHm_2ChTwMy_3$

   Have your instructor check your answer, and initial here: _____

Exercising Your Knowledge

8.   How many sandwiches are made in the equation above?

9.   If you don't want any leftover ingredients, why must you make more than one sandwich?

10.   Going back to the original grilled cheese sandwich, if a slice of bread has a mass of 25 grams, and a slice of cheese has a mass of 20 grams, what is the mass of the grilled cheese sandwich?

11.   How did you figure out this answer?

12.   Balance each of the chemical equations below as you did the sandwich equation, using whole numbers to represent the numbers of each particle that make the numbers of atoms of each element equal on both sides of the equation. (Helpful hints: it is useful to start balancing by starting with the chemical formula with the largest number of atoms, and balancing may start with either a reactant or a product.)

      a.   ___ H_2 + ___ O_2 → ___ H_2O

      b.   ___ HNO_3 + ___ Cu → ___ $Cu(NO_3)_2$ + ___ H_2

      c.   ___ Fe + ___ O_2 → ___ Fe_2O_3

      d.   ___ CH_4 + ___ O_2 → ___ CO_2 + ___ H_2O

      e.   ___ $NaNO_3$ → ___ $NaNO_2$ + ___ O_2

Summarizing Your Thoughts

13.   In balancing an equation, why is it advisable to start with the chemical formula with the largest number of atoms?

14.   The (unbalanced) chemical equation representing water decomposing into its component elements is: H_2O → H_2 + O_2. Write out this chemical equation, and balance it. How is this equation related to the one in 12a above?

15.   A (seemingly) easier but incorrect way to do this would be to change the equation to be: H_2O_2 → H_2 + O_2. Why is changing coefficients to balance a chemical equation acceptable, but changing subscripts to balance a chemical equation unacceptable?

16. The Law of Conservation of Mass states that in a chemical reaction mass is not created or destroyed, but instead changes form. How is this demonstrated in making a grilled cheese sandwich?

17. In a balanced chemical equation what must be equal on both sides of the equation, the sum of the coefficients, the number of molecules, or the number of atoms of each element?

18. How does a balanced chemical equation demonstrate the Law of Conservation of Mass?

19. As a group, come up with three concepts communicated in today's work to which you will need to dedicate additional study time outside of class in order to become more comfortable with them.

ACTIVITY 10.2

Objectives

- Understand what it means for an equation to be balanced

- Be able to identify when an equation is balanced

Getting Started

We ask that you find some macroscopic object to represent an atom. Your macroscopic item could be small bingo chips, candies (M&Ms™ or Skittles™), or even wadded up pieces of paper. However, what is important as you begin is that you prepare a color-code key so that you can remember which atoms are being represented.

Suggested Color Scheme		**Color Scheme Used**
Oxygen:	red	Oxygen:
Sodium:	purple	Sodium:
Chlorine:	green	Chlorine:
Calcium:	orange	Calcium:

The Model

Chemical reactions may be described in words or in a *chemical equation*, which is a representation of a reaction using chemical symbols and chemical formulas. One chemical reaction could be described in words as calcium oxide reacts with sodium chloride in a double-replacement reaction to form calcium chloride and sodium oxide.

Reviewing the Model

1.   What are the products in the chemical reaction written above?

2.   What type of reaction is written above?

Exploring the Model

3.   If the products were not stated in the description of the chemical reaction above, could you determine what those products would be? How?

4.   To represent a reaction as a chemical equation, each component must be represented as a chemical formula. Fill in the blanks below with the proper chemical formula for each compound involved in this reaction. Be sure to use the proper number of anions and cations to make the overall charge of each compound zero.

   _____   +   _____   →   _____   +   _____

5. Model this reaction using your chips. Start with one of each of the reactant particles in the proper grouping in the reactants box on the paper, and move the chips into the product box as product compounds are formed.

    a. Were you able to form all the products indicated in the chemical equation?

    b. If yes, how were you able to do it? If no, why were you not able to complete the task, and what else is needed to form the products indicated?

6. In a chemical equation, coefficients are numbers in front of chemical formulas that indicate how many particles of each compound are required for the reaction to take place as written. In the blanks below, place the chemical formulas for each component in the chemical equation, and indicate before each chemical formula the number of particles (or groups) needed for the reaction to take place (the first one is done for you)

    $\underline{\quad 1 \quad}$ $\underline{\quad CaO \quad}$ $+$ $\underline{\quad\quad}$ $\underline{\quad NaCl \quad}$ $\rightarrow$ $\underline{\quad\quad}$ $\underline{\quad\quad\quad\quad}$ $+$ $\underline{\quad\quad\quad\quad}$

7. Model this reaction again, this time using the number of each particle indicated in the chemical equation above. Were you now able to form all the products indicated in the chemical equation?

Exercising Your Knowledge

A chemical equation is said to be balanced when whole numbers of each product molecule may be changed into whole numbers of each reactant molecule (with the numbers of each compound indicated by coefficients, as you did above).

8. Below are a series of chemical equations. Circle each chemical equation below that is **not** properly balanced.

    a. $CaCO_3 \rightarrow CaO + CO_2$

    b. $2\,NiCl_2 + NaOH \rightarrow Ni(OH)_2 + 2\,NaCl$

    c. $Ag + Cl_2 \rightarrow AgCl$

    d. $NaOH + HNO_3 \rightarrow H_2O + NaNO_3$

    e. $C_2H_2 + 2\,O_2 \rightarrow 2\,CO_2 + 2\,H_2O$

Summarizing Your Thoughts

9. How can you determine from a written chemical equation if a chemical equation is balanced?

10. What does it mean if a chemical symbol in a chemical formula does not have a subscript?

11. What does it mean if a chemical formula in a chemical equation does not have a coefficient?

12. In a balanced chemical equation, what can be said about the number of atoms of each element in the products and reactants?

13. In a balanced chemical equation, what can be said about the total mass of the reactant particles and the total mass of the product particles? Can the same be said about an unbalanced chemical equation?

14. Were there any parts of this activity that were difficult for some members of the group to understand? If so, how did your group work together to make sure the material was clear for all of the group members?

ACTIVITY 10.3

Objective

- Understand how coefficients in a balanced chemical equation communicate information about the quantities of compounds involved in a reaction

The Model

$$MgCO_3 + 2\ LiF \rightarrow MgF_2 + Li_2CO_3$$

Reviewing the Model

1.   How many different compounds are shown in the Model?

2.   Which compound has a coefficient of 2 in this equation?

Exploring the Model

3.   What does the coefficient of 2 mean in the preceding equation?

4.   If one unit of magnesium carbonate reacts as shown in the Model, how many lithium fluoride units are needed?

5.   How many units of each product will be formed in the reaction if one unit of magnesium carbonate reacts?

6.   If ten units of magnesium carbonate react, how many lithium fluoride units will have reacted with them?

7.   How do you know this is true?

8.   If one million units of magnesium carbonate react, how many lithium fluoride units will have reacted with them?

9.   If 0.8 moles of magnesium carbonate react, how many moles of lithium fluoride will have reacted with them?

10.   The coefficients present in a balanced chemical equation can be used as conversion factors to calculate how many units of any compound will be formed or used in a chemical reaction. In this case, we can think of 1 unit of $MgCO_3$ = 2 units of LiF (where a unit is any countable number, usually an individual molecule or a mole. Use this relationship to set up a conversion factor that shows how to do the math in problem #9:

   0.8 moles $MgCO_3$ x _____   = ___ moles LiF

Exercising Your Knowledge

11.   For the following chemical equation, balance the equation, and use the proper ratios of molecules to determine how much aluminum chloride would be formed in this reaction if 100 atoms of aluminum were consumed in the reaction:

   _____ Al + _____ Cl_2 → _____ $AlCl_3$

   100 atoms Al x _____ = _____ molecules $AlCl_3$

12.   How many chlorine molecules would be consumed in this reaction if 100 atoms of aluminum were consumed?

   100 atoms Al x _____ = _____ molecules Cl_2

13.   Magnesium hydroxide is used as an antacid in milk of magnesia and reacts with hydrogen chloride in the stomach to form water and magnesium chloride. Write out the chemical equation for this reaction with the correct chemical formulas, and balance the equation.

14.   If a dose of milk of magnesia contains 0.020 moles of magnesium hydroxide, how much water will be formed when it reacts?

Summarizing Your Thoughts

15.   List counting units that would be acceptable for use in the way you think about a chemical equation.

16.   In chemistry we use the mole most often to represent the amount of a substance. Why are moles used rather than individual molecules?

17.   What information does a balanced chemical equation give that an unbalanced chemical equation does not?

18.   In determining how much of a compound is produced in a reaction, why is it important to use a balanced chemical equation?

19.   When something was not well understood in this activity, which group member was best able to explain it to the other members of the group? How did they explain the concepts so that the other members of the group understood them better?

20.   With this in mind, what can other members of the group do to better understand the material presented, and what study techniques will be most useful in preparing for the examination over this material?

CHAPTER 11

What's Going On? (Chemical Reactions)

Chapter Goals

- Understand what chemical reactions are and how chemical equations represent chemical reactions

- Be able to define reactants and products in a chemical reaction

- Understand why coefficients are used in chemical equations

- Be able to identify combination, decomposition, single-replacement, double-replacement, and combustion reactions

Getting Started

Things are always happening in a chemical reaction. Observations that make it appear nothing is happening are misleading. Nothing is static in the chemical world. The dynamic nature of chemistry is a big challenge for a textbook. How do we show change when our words and pictures can't change? Well, in this chapter we will ask you to help.

We ask that you find some macroscopic object to represent an atom. Your macroscopic item could be small bingo chips, candies (M&Ms™ or Skittles™), or even wadded up pieces of paper. However, what is important as you begin is that you prepare a color-code key so that you can remember which atoms are being represented.

Have a little fun and try to gain an understanding the dynamic nature of chemistry.

ACTIVITY 11.1

Objectives

- Understand how chemical equations relate to a particulate-level process

- Be able to define and identify reactants and products in a chemical reaction

Getting Started

Chemical reactions are the rearrangement or recombination of the starting material's atoms or molecules (the reactants) into new substances (the products). Observations of chemical reactions can be made on the macroscopic level by visual inspection of a chemical reaction. However, a key aspect to chemistry is the ability to visualize, link, and represent what happens on the particulate level (the individual molecules or atoms) in chemical equations to observations that are made on the macroscopic level.

A symbolic representation of a chemical reaction that uses chemical symbols to represent atoms, elements or compounds and the changes they undergo is used throughout science. But the best chemists will look at the symbolic representation and think about the nanoscale changes that are occurring.

The Model

Below is a *chemical equation*, a symbolic representation of a *chemical reaction*, that uses chemical symbols to represent molecules and the changes they undergo. A particulate-level representation of this reaction is shown in the boxes below:

$$C + O_2 \rightarrow CO_2$$

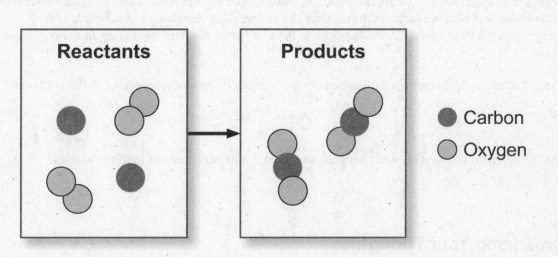

Reviewing the Model

1. Give the symbol and name for all the elements represented in the model.

2. Give the symbol and name for all the compounds represented in the model.

Exploring the Model

3.   How are atoms and molecules represented differently in this model?

4.   Describe the macroscopic properties of each element or compound in the model.

5.   In the model, place a chip representing carbon over the black circles in the box on the left and chips representing oxygen over the other circles in the box on the left. Move the chips to the box on the right, and re-arrange them to form CO_2. After you go through the process outlined above, how many chips are left in the reactant box?

6.   Did the number of chips required change as you proceeded? _____

7.   How do your observations about the macroscopic changes relate to the changes you observe in the model?

Exercising Your Knowledge

8.   Imagine a collection of a lot of carbon atoms and oxygen molecules in an airtight container that recombine and become carbon dioxide over the course of several hours. As the reaction progresses (as time passes), what happens to the number of reactant molecules in the container?

9.   As the reaction progresses, what happens to the number of product molecules in the container?

10.   As the reaction progresses, what happens to the total number of atoms in the container?

Summarizing Your Thoughts

11.   In your own words, what is the definition of a reactant in a chemical reaction?

12.   In your own words, what is the definition of a product in a chemical reaction?

13. How is the relationship between reactants and products shown symbolically in a chemical equation?

14. From your observations, explain why the statement, "Matter can't be created or destroyed in a normal chemical reaction," is true.

15. Describe any additional questions about chemical equations that came up in your discussion of this activity. What possible answers to those questions did your group discuss?

ACTIVITY 11.2

Objectives:

- Understand how atoms rearrange in combination and decomposition reactions

- Understand what coefficients represent in a chemical equation

Getting Started

In Activity 11.1 we looked at a simplified model reaction. For this activity, it will be easiest for you to use the given chip colors to represent each element.

Model

The following is the suggested system used throughout chemistry for coloring models of atoms:

Suggested Color Scheme		**Color Scheme Used**
Carbon:	black	Carbon:
Oxygen:	red	Oxygen:
Chlorine:	green	Chlorine:
Nitrogen:	blue	Nitrogen:
Hydrogen:	white	Hydrogen:

Reaction 1: $CO + Cl_2 \rightarrow COCl_2$

Reaction 2: $N_2H_4 \rightarrow N_2 + 2 H_2$

Reviewing the Model:

1. What are the names for all the elements represented by chemical symbols in Reactions 1 and 2?

2. Identify all the compounds in the model.

3. What is the name of the reactant compound in Reaction 1? _____

4. Sketch and label a nanoscale view of each molecule in Reaction 1 and 2.

Exploring the Model

5. Draw two boxes on a sheet of loose paper (you will not turn this page in). Label one box "before" and the other "after." Use your chips (or candies) to create the "before" picture for Reaction 1.

   a. How many atoms are in the "before" box? _____

   b. How many molecules are in the "before" box? _____

6. Rearrange the chips from your Reaction 1 "before" box as you move them to the "after" box.

   a. How many atoms are in the "after" box? _____

   b. How many molecules are in the "after" box? _____

7. Use your chips (or candies) to create the "before" picture for Reaction 2.

   a. How many atoms are in the "before" box? _____

   b. How many molecules are in the "before" box? _____

8. Rearrange the chips for Reaction 2 as you move them to the "after" box.

   a. How many atoms are in the "after" box? _____

   b. How many molecules are in the "after" box? _____

Exercising Your Knowledge

9. In this reaction, a "2" is placed in front of one of the product molecules in the chemical equation. What does this number represent?

10. Based on your observations, explain why the coefficient "2" was necessary in the chemical equation.

11. Two types of reactions are represented in this activity: combination and decomposition. For each reaction, decide whether or not the reaction is a combination reaction. Then determine whether the equation shown needs to include any coefficients. If coefficients are needed, write them in the appropriate place.

	Combination, decomposition, or neither?	Coefficients needed?
$Hg + S \rightarrow HgS$	_____	_____
$Al + Cl_2 \rightarrow AlCl_3$	_____	_____
$HgO \rightarrow Hg + O_2$	_____	_____
$NH_3 + I_2 \rightarrow NI_3 + H_2$	_____	_____

Summarizing Your Thoughts:

12. In your own words, give the definition of a combination reaction. Check this definition with your instructor.

13. If a chemical equation were given, what characteristics would you use to determine if the equation represented a combination reaction?

14. Another type of reaction, a decomposition reaction, could be described by the chemical equation $H_2CO_3 \rightarrow H_2O + CO_2$. In your own words, give the definition of a decomposition reaction.

15. In grammatically correct sentences, write a paragraph to compare and contrast decomposition reactions and combination reactions.

16. What is the difference between a chemical reaction and a chemical equation?

17. Briefly describe why having a standardized color scheme would be helpful to a chemist.

18. Write a short paragraph that explains why coefficients are needed when writing chemical equations.

ACTIVITY 11.3

Objectives

- Understand how atoms rearrange in single-replacement reactions

- Understand which subatomic particles are involved in single-displacement reactions

- Associate the terms *oxidation* and *reduction* with the transfer of a subatomic particle

Getting Started

Use your chips to help you see that we can replace one atom from a compound, but a new twist has been introduced. You will need to recall information about how to determine the ionic charge on a cation to complete this exercise. Look back to your earlier summaries to see how to determine the charge on a cation.

The Model

Model this reaction with all the necessary reactant particles in the proper grouping in the "reactants" box on the paper, and rearrange the chips to form all the product particles in the "products" box on the paper using the chip colors indicated below for each atom of that element.

Suggested Color Scheme		Color Scheme Used
Chlorine:	green	Chlorine:
Iron:	amber	Iron:
Cobalt:	pink	Cobalt:

Reaction 3: $Fe + CoCl_2 \rightarrow Co + FeCl_2$

Reviewing the Model

1. How many different chips (elements) are in the "reactants" box before the reaction occurs? _____

2. How many different chips (elements) are in the "products" box after the rearrangement? _____

3. Did the identity of any element change? _____

4. Did the total number of atoms change? _____

Exploring the Model

5. Why are there two chloride ions paired with one cobalt ion in the "reactants" box?

6. What happens to the chloride ions during this reaction?

7. In a chemical equation, all components must have an overall charge of zero. With this in mind, what is the ionic charge on iron as a reactant? _____

8. What is the ionic charge on the iron ion as a product? _____

9. Does the charge on cobalt change as the reaction proceeds? _____

Exercising Your Knowledge

10. For iron to have two different ionic charges, what subatomic particle (proton, neutron, or electron) is involved? _____

11. Did the iron gain or lose the particle(s)? _____

12. How many particles were involved in the change? _____

13. Where did the particle(s) go?

14. Oxidation occurs when an atom or ion loses electrons through a chemical reaction. In the chemical equation above, which particle is oxidized? _____

Summarizing Your Thoughts

15. Reaction three represents a single-replacement reaction. Single-replacement reactions commonly occur with ionic compounds and metals. In your own words, what has to happen during the course of a single-replacement reaction?

16. What clues are present in a chemical equation that will help you recognize single-replacement reactions?

17. Oxidation and reduction are opposite processes that must always occur together. In the reaction above, the cobalt ion is reduced. Based on this information, give a definition of the term reduction.

18. Why must oxidation and reduction always occur together?

ACTIVITY 11.4

Objective

- Recognize trends in double-replacement reactions in order to be able to predict the identity of possible products

Getting Started

Things are getting a little more complicated, but there are definite trends in this model. You should recall your nomenclature rules and be prepared to recognize some polyatomic ions in order to see these trends.

The Model

Suggested Color Scheme	Color Scheme Used
Sodium: purple	Sodium:
Calcium: orange	Calcium:

Reaction 4: $2\ NaOH + CaCl_2 \rightarrow Ca(OH)_2 + 2\ NaCl$

Reviewing the Model

1.   What are the names of the reactant compounds in this reaction?

2.   How many different colored chips are used for this reaction? _____

3.   Identify the polyatomic ion in the reaction. _____

Exploring the Model

4.   Model the reactant molecules in the "reactants" box using the chips or candies.

5.   Group the chips to form the product compounds in the "products" box. When the reaction is complete, how many chips are in the "products" box? _____

6.   Circle the chemical symbols for the cations in the chemical equation. Write the names of the cations present in this reaction.

$$2\ NaOH + CaCl_2 \rightarrow Ca(OH)_2 + 2\ NaCl$$

7.   Does the ionic charge on the ions change as the reaction proceeds? _____

8.   Why is it unreasonable to predict that one of the products would be CaNa?

9.   What are the reasonable combinations for the sodium cation given only those ions available as reactants?

10. In a complete sentence, what happens to the complex ion during the course of the reaction? Does it break apart?

11. Why is the "2" needed before NaCl in this equation?

Exercising Your Thoughts

12. Write the chemical equation, including coefficients if necessary, for the reaction of calcium hydroxide with iron(II) sulfide to form iron(II) hydroxide and calcium sulfide.

13. Use your chips to model the chemical change in the iron(II) sulfide reaction above; then sketch what the "products" box looks like after the reaction is complete (put the chemical symbol for each atom inside its corresponding circle).

14. In a reaction between calcium chloride and potassium carbonate, what products would be formed if the model were used as guidance?

Summarizing your Thoughts

15. The model represents a double-replacement reaction. In your own words, give the definition of a double-replacement reaction.

16. In some reactions, polyatomic ions (such as OH⁻ in the reactions above) can be treated as a group in balancing a chemical equation rather than being treated as separate atoms (O and H). Why might it be beneficial to work with these atoms as a group rather than as individual atoms?

17. Review your exercises, and write a list of clues to look for when trying to predict the types of products that are produced in a double-replacement reaction.

18. Below are four general equations for combination, decomposition, single-replacement and double-replacement reactions with letters representing different elements. In the blank following each chemical equation, state what type of chemical reaction it best represents.

$AB + C \rightarrow CB + A$ _____

$QR \rightarrow Q + R$ _____

$DE + FG \rightarrow DG + FE$ _____

$X + Y \rightarrow XY$ _____

19. There are several possible "jobs" for group members in this activity. Write down the name of each member of the group and what job they did to help the group complete this activity.

ACTIVITY 11.5

Objectives

- Understand how atoms rearrange in combustion reactions
- Link macroscopic to particulate levels

Getting Started

This reaction describes the burning of methane (the principal component of natural gas). Methane has a series of compounds call hydrocarbons that behave very similarly. Use your chips (candies) to explore this class of chemical reactions.

The Model

A series of combustion reactions:

$$\text{Reaction 5: } CH_4 + 2\,O_2 \rightarrow CO_2 + 2\,H_2O$$

$$\text{Reaction 6: } 2\,C_2H_6 + 7\,O_2 \rightarrow 4\,CO_2 + 6\,H_2O$$

$$\text{Reaction 7: } 2\,C_6H_6 + 15\,O_2 \rightarrow 12\,CO_2 + 6\,H_2O$$

Reviewing the Model

1. Which element above is presented as a diatomic molecule? _____

2. The hydrocarbons are listed first in each equation. What elements are always present in a hydrocarbon? _____

3. When a hydrocarbon combusts (burns), what does it react with from the air? _____

4. What compounds are produced in all combustion reactions? _____   _____

Exploring the Model

5. The "2" in front of O_2 in Reaction 5 in this chemical equation is called a coefficient. What does a coefficient represent?

6. Why are the coefficients not the same for CO_2 in these reactions?

7. Use your chips to model Reaction 5. Now try modeling this reaction with one oxygen molecule instead of two in the "reactants" box. Can the proper number of product molecules be obtained? _____

Exercising Your Knowledge

8. What is another hydrocarbon that is commonly used as fuel? _____

9. Another hydrocarbon that burns is ethene, C_2H_4. Obtain enough chips to produce CO_2 and H_2O with no atoms left over. How many reactant oxygen molecules are needed for this to occur? _____

10. Write a balanced chemical equation for the combustion of ethene.

11. What about the chemical equations describing the combustion of ethene and methane is the same?

12. What two things about these chemical equations are different?

13. Hydrogen and oxygen gases may react to form water as described by the chemical equation: $2\,H_2 + O_2 \rightarrow H_2O$. Would you consider this a combustion reaction? Why or why not?

Summarizing Your Thoughts

14. Summarize the things you should be able to recognize in order to predict when a reaction is a combustion reaction.

15. If you were given only the formula for a hydrocarbon, describe what other compounds or elements would be needed to predict the correct chemical equation for the combustion of that hydrocarbon.

16. Once you predict the reaction, you need to assure that the number of each element is the same on the reactant and product sides. Describe the steps you will take to assure that this equality is true.

17. One member of your group gets a gold star for their work today. Decide as a group which member deserves the gold star. Why did that person get the gold star today?

18. For each other member of the group, state how that member can improve their participation on these activities to deserve the gold star in a future activity.

CHAPTER 12

How Much Is Possible? (Limiting Reactants)

Chapter Goals

- Understand and be able to define limiting and excess reactants in a chemical process
- Be able to identify limiting and excess reactants given molar or mass amounts of reactants
- Be able to determine percent yield in a chemical reaction

ACTIVITY 12.1

Objectives

- Understand what limits products
- Be able to define the terms *excess reactant* and *limiting reactant*

The Model

Making a grilled cheese sandwich can be represented by the chemical equation:

$$2 \text{ Bd} + \text{Ch} \rightarrow \text{Bd}_2\text{Ch}$$

Reviewing the Model

1.   We have made up symbols for the "reactants" needed to make a grilled cheese sandwich. What do the symbols "Bd" and "Ch" represent in this chemical equation?

2.   What does the '2' represent at the beginning of the above chemical equation?

3.   What does "Bd$_2$Ch" represent on the right side of the preceding equation?

Exploring the Model

4.   You are given a loaf of bread that contains 28 slices of bread and a package of cheese that contains 16 slices of cheese. How many sandwiches could you make?

5.   Use complete sentences to explain how you figured out how many sandwiches you could make.

6.   What will be left over when all the sandwiches are made?

Exercising your Knowledge

7.   At a tricycle factory, a shipment with 400 seats and 600 wheels comes in. Assuming that all other parts are available, how many tricycles can be made?

8.   In the tricycle example, what information did you know that wasn't stated in the problem, but was required to solve the problem?

9.   There are more wheels than there are seats, so how can there be seats left over after all of the tricycles are assembled?

10.   Write out the balanced chemical equation describing a double-replacement reaction between calcium bromide and lithium hydroxide.

11.   If ten moles each of calcium bromide and lithium hydroxide are combined and react completely according to your balanced chemical equation, how many moles of lithium bromide will be formed?

Summarizing your Thoughts

12.   In the preparation of cheese sandwiches with a limited amount of bread, the bread could be referred to as the limiting reactant. In your own words, give a definition for the term "limiting reactant."

13.   When two reactants are present and one is the limiting reactant, the other is the excess reactant. In your own words, give a definition for the term "excess reactant."

14.   Describe how the balanced chemical equation is used as part of your calculation of the maximum amount of product you can make.

ACTIVITY 12.2

Objectives

- Be able to determine the limiting reactant in a chemical reaction

- Be able to determine how much of a compound may be produced in a limiting reactant problem

The Model

Gaseous oxygen (O_2) and gaseous hydrogen (H_2) combine to form water (H_2O)

Reviewing the Model

1.   What are the reactants in this reaction?

2.   What is the product in this reaction?

3.   Write out the balanced chemical equation for the reaction in the Model.

4.   What is the ratio of molecules of oxygen consumed in the reaction to molecules of hydrogen consumed?

Exploring the Model

Prepare a new sheet of paper by drawing two boxes. Label these boxes "before" and "after." Use bingo chips or candies (Skittles™ or M&Ms™) to model the reaction. For the activity it will be easiest for you to use the red chips to represent oxygen and the white chips to represent hydrogen. This is the standard system for coloring models of atoms that is used throughout chemistry, but if you can't find white candies make an appropriate substitution.

1.   Start with all the necessary reactants as described by your balanced equation in the proper grouping in the "before" box on the paper, and rearrange the chips to form all of the product particles in the "after" box on the paper. Describe your observations.

2.   Repeat the reaction by modeling this reaction with two oxygen molecules and two hydrogen molecules. Create as many water molecules as possible. How much of each reactant is left after the reaction is complete?

Oxygen –

Hydrogen –

Water –

3.   If two dozen oxygen molecules were combined with two dozen hydrogen molecules, what would be the limiting reactant, and how much water could be formed?

4. If two moles of oxygen molecules were combined with two moles of hydrogen molecules, what would be the limiting reactant, and how much water could be formed?

5. Explain how scaling the number of particles from two dozen to two moles changes the way you think about how many water molecules could be formed.

Exercising your Knowledge

6. Write the balanced chemical equation describing a double-replacement reaction between sodium nitrate and calcium carbonate.

7. If 1.43 moles of sodium nitrate react with an excess amount of calcium carbonate, how much sodium carbonate would be formed? The framework is set up below to use the ratios from the balanced chemical equation:

8. 43 moles $NaNO_3$ x _____ = _____ moles Na_2CO_3

9. If 0.87 moles of calcium carbonate react with an excess amount of sodium nitrate, how much sodium carbonate will be formed?

10. If 1.43 moles of sodium nitrate and 0.87 moles of calcium carbonate are combined and react as shown, how much sodium carbonate will be formed? What is the limiting reactant in this example?

11. Below are four reactions that describe the combustion of four different hydrocarbons.

$CH_4 + 2 O_2 \rightarrow CO_2 + 2 H_2O$

$2 C_2H_6 + 7 O_2 \rightarrow 4 CO_2 + 6 H_2O$

$C_2H_4 + 3 O_2 \rightarrow 2 CO_2 + 2 H_2O$

$2 C_2H_2 + 5 O_2 \rightarrow 4 CO_2 + 2 H_2O$

In a container, 2.0 moles of one of these hydrocarbons combines with 6.0 moles of oxygen and is burned. After the reaction is complete, some hydrocarbon is left in the tank. Which of the above reactions could it be? How did you solve this problem?

Summarizing your Thoughts

12. Given molar quantities of two reactants in a chemical reaction, how would you determine which is the limiting reactant?

13. How would you determine the quantity of a product that may be formed in a limiting reactant problem?

14. Explain why we did not use gram quantities to make our determination of the limiting reactant.

15. If mass quantities (in grams) were given, how would you determine which is the limiting reactant?

ACTIVITY 12.3

Objective

- Understand what "percent yield" represents and be able to calculate percent yield for a chemical reaction

The Model

A friend comes over to visit and decides to make grilled cheese sandwiches. In your kitchen, there are 16 slices of bread and 8 slices of cheese, which he uses up in making the sandwiches, and he emerges from the kitchen with 6 sandwiches on a plate.

$$\% \, yield = \frac{actual}{expected} \times 100$$

Reviewing the Model

1. How many sandwiches did you expect?

2. How many sandwiches were prepared?

Exploring the Model

3. Using the equation in the Model, what is the percent yield of the sandwiches from the starting material?

4. Just as any good scientist should do, hypothesize about what may have happened to the missing sandwiches.

5. List some other nouns that could be used for the expected amount.

6. Which of these clauses might be the same as an actual amount?

    a.   35 g of lead nitrate was collected

    b.   14 g of NaCl was reacted

    c.   2 g of sugar was added

    d.   8.8 g of iron oxide was obtained

Exercising your Knowledge

7.   A sample of 58.44 grams of sodium chloride reacts with excess silver nitrate in a double-replacement reaction, and silver chloride is collected. How much silver chloride would be expected to form in this reaction? (Don't forget to think in terms of moles.)

8.   If 103.24 grams of silver chloride are actually obtained, what is the percent yield?

9.   In a reaction, 25 grams of sodium hydroxide reacts with hydrogen chloride to form water and 25 grams of sodium chloride. Is the percent yield in this process 100%? Describe your thought process.

10.   Acetone is a widely used chemical with the chemical formula C_3H_6O. It is commonly made from isopropyl alcohol, C_3H_8O, by removing molecular hydrogen, H_2. A new process for production of acetone is developed, and in order for it to compete with current processes, at least 92% of the expected product must be obtained. If 323 moles of isopropyl alcohol are used in the system, what is the minimum number of moles of acetone that must be obtained for the new process to be competitive?

Summarizing your Thoughts

11.   The equation used to calculate percent yield was given in the Model. Write a definition in your own words for the term "percent yield."

12.   Explain why you can't use grams of product divided by grams of reactant to calculate percent yield.

13. Provide at least two reasons you think percent yield might be important to a chemist.

14. Which aspect of this activity was most difficult for your group to complete?

15. What did your group do in order to get around this difficulty?

16. List each member of the group. Have each member give two skills that they need to develop (or concepts that they need to work on) to master the material presented for the next exam.

CHAPTER 13

Gases and the Laws

Chapter Goals

- Explain or predict physical phenomena relating to gases in terms of the ideal gas model

- Understand how pressure influences volume (Boyle's law)

- Understand how temperature influences volume (Charles's law)

- Understand how temperature and pressure combine to influence volume (Combined gas law)

- State the values associated with standard temperature and pressure (STP) for gases

ACTIVITY 13.1

Objective

• Explain or predict physical phenomena relating to gases in terms of the ideal gas model

The Model

Imagine that you and a friend have been stranded on a small raft in the middle of the ocean for days.

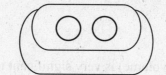

Figure 13.1 You and a friend in an oval-shaped raft

Reviewing the Model

1. Given your current surroundings and situation, do your respective physical sizes (or volumes) have any influence on each other?

2. What is the major factor that allows you to interact with your friend?

Exploring The Model

Imagine that you and your friend get *really* tired of interacting with one another and choose to inflate the extra raft. Your friend climbs into the second raft, and the ocean currents set you drifting in opposite directions.

Your friend is here

You are here

Figure 13.2

3. Given your current surroundings and situation, does your physical size (or volume) have any influence on your friend, who is now many, many miles away?

4. If you do not possess any communication devices, what is the major factor preventing you from interacting with your friend or the nearest human?

Exercising Your Knowledge

5. Describe **why** your physical size (or volume) is very significant to you and your friend while in the raft together.

6. Describe **why** your physical size (or volume) when you are alone way out in the middle of the ocean is insignificant to your friend or anyone else.

Summarizing Your Thoughts

7. Recalling that gas particles spread out to fill their container, how would an individual gas particle's physical size (or volume) relate to a sample of gas particles?

ACTIVITY 13.2

Objective

- Understand how pressure influences the volume a gas may occupy (Boyle's law)

Getting Started

If you squeeze a balloon with your hands, it will pop if the pressure around the balloon decreases and it expands to the point of popping. You do this by applying a force (your hands squeezing) per unit area (the surface area of your hand in contact with the balloon). In other words, pressure is equal to force per unit area $(P = F/A)$.

The Model

Figure 13.3

Consider the container (box) containing two gas particles (black dots) in Figure 13.3 in answering the following questions.

Reviewing the Model

1. In grammatically correct sentence(s) describe the size of the gas particles relative to the size of the container.

2. In grammatically correct sentences describe how the size (or volume) of one gas particle influences the other gas particle.

3. At their current locations in the container are the two gas particles forced to interact with one another?

Exploring The Model

Figure 13.4

Consider the container (box) containing two gas particles (black dots) in Figure 13.4 as the same container as in Figure 13.3 but after experiencing a pressure that reduced its size considerably.

4. In grammatically correct sentence(s) describe the size of the gas particles relative to the size of the container.

5. Describe how the size (or volume) of one gas particle influences the other gas particle.

Exercising Your Knowledge

6. Explain **why** the two gas particles are forced to interact with one another.

7. Explain **the way in which** the two gas particles are forced to interact with one another.

Summarizing Your Thoughts

8. If the container that holds the gases experiences an external pressure strong enough to force the gas particles so close together that they must interact with one another, what has effectively been removed completely from this sample?

ACTIVITY 13.3

Getting Started

The greater the amount of water that surrounds an object, whether it be a living being or a balloon, the greater the pressure that will be exerted on that object. Imagine what happens to gas particles in a balloon as a filled balloon is taken deeper and deeper below the surface of the ocean.

The Model

Figure 13.5

Reviewing the Model

1. What happens to the balloon when it goes underwater?

2. If the pressure exerted by the water is doubled, what effect will this have on the volume of the balloon?

Exploring the Model

3. Since the number of gas particles in the balloon does not change (none can escape), what does this mean for the gas particles within the balloon?

4. Describe **what** would happen if you took an empty balloon to the bottom of the ocean, filled it with air and sealed it and then returned it to the surface?

5. Describe from a gas-particle point of view **why** the scenario in Question 3 above would happen as you answered.

Exercising Your Knowledge

Volume and pressure are inversely proportional; $P \propto 1/V$. In other words, as an external pressure increases (becomes larger), the volume occupied by a fixed number of gas particles at constant temperature decreases (becomes smaller). So pressure and volume can contribute in different proportions that always equal a constant value; $P \times V = k$. An initial and final set of circumstances can be equated to one another through this constant value.

$$P_i \times V_i = P_f \times V_f$$

6. Given that doubling the pressure reduces a volume of gas to half its original size, fill in the following table:

Table 13.1

Initial Pressure (P_i), atm	Initial Volume (V_i), L	$P_i \times V_i =$ atm ·L	Final Pressure (P_f), atm	Final Volume (V_f), L	$P_f \times V_f =$ atm ·L
1	10		2	5	
	4	4		2	4
2			1		12

Summarizing Your Thoughts

7.   If the container that holds the gases experiences an external pressure strong enough to force the gas particles so close together that they must interact with each other, how might this change their physical state (e.g., gas)?

8.   What states are possible, and to what degree are the particles forced to interact with one another for each state?

ACTIVITY 13.4

Objective

•   Understand how temperature influences the volume that a gas may occupy (Charles's law)

Getting Started

Imagine you have just purchased a bouquet of balloons for your New Year's Eve party from the store. You place them in the car and then stop by the grocery store for some cake and ice cream (and maybe a few other items). When you return, your balloons don't look so cheerful. The cool January air had done a trick on them.

In this activity, we will explore why the volume of the balloon changed. Remember to think about how all of those teeny, tiny gas particles (atoms or molecules) as you proceed, and see if you can't return to your party with balloons that resemble those that you purchased.

The Model

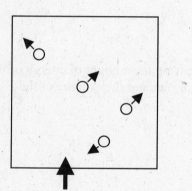
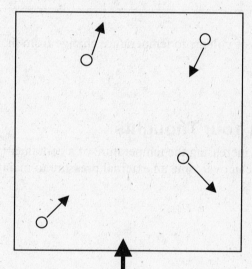

Temperature 1   Twice Temperature 1

Figure 13.6

Exploring the Model

1.   In grammatically correct sentences, describe **how** changing the temperature changes the size of the container in the model.

2.   In grammatically correct sentences, describe what else changes about the model.

Exercising Your Knowledge

3.   Given that doubling the temperature doubles the original volume a gas may occupy, fill in the following table:

Table 13.2

Experiment	Initial Temperature (T_i)	Initial Volume (V_i)	$V_i/T_i =$	Final Temperature (T_f)	Final Volume (V_f)	$V_f/T_f =$
1	100	10		200	20	
2	200		0.02	400		0.02
3		8		100		0.02

4.   Does the ratio of volume to temperature change from the initial condition to the final conditions?

Summarizing Your Thoughts

5.   Describe how increasing the temperature of a container (increasing the amount of energy of the gas particles) can overcome an external pressure to maintain or increase the volume of the container.

6.   If the ratio is a constant, could we set the two ratios equal to each other? Using the column headings from Table 13.2, write an equality that is true.

ACTIVITY 13.5

Objective

- Understand how temperature and pressure combine to influence volume (Combined gas law)

The Model

Consider the phase diagram below and what happens to gas particles as pressure, temperature, or both pressure and temperature change.

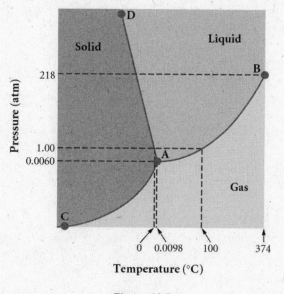

Figure 13.7

Exercising Your Knowledge

1. In grammatically correct sentences describe what is happening to individual gas particles as temperature increases. Assume a constant pressure.

2. In grammatically correct sentences describe what is happening to individual gas particles as pressure increases. Assume a constant temperature.

3. In grammatically correct sentences describe what is happening to individual gas particles as pressure and temperature increase at the same time.

Summarizing Your Thoughts

4. Describe how increasing the temperature of a container while increasing the pressure on that container could completely cancel any change in that container's volume.